Gardening
Nutrition

An Oxfam document compiled by
Arnold Pacey

Intermediate Technology Publications

Published by Intermediate Technology Publications Ltd
103/105 Southampton Row, London WC1B 4HH, UK

First published February 1978
Second impression June 1979
Third impression September 1980
Fourth impression March 1982
Fifth impression October 1983
Sixth impression November 1984
Seventh impression June 1988

ISBN 0 903031 50 7

Printed by The Russell Press Ltd, Nottingham

Contents

List of Illustrations

Acknowledgements

Apart from the accounts of practical experience gleaned from these field reports, and information obtained from books listed in the bibliography, the compiler of this manual has benefited from the advice of members of the Intermediate Technology Development Group Nutrition Panel, and from discussions with Dr G.A.C. Herklots and with many Oxfam field workers. Thanks are also due to the Food and Agriculture Organization of the United Nations (F.A.O.) for permission to reproduce Figures 5-8 from the booklet, *Market Gardening,* in their 'Better Farming Series'.

Preface

This is the second in a series of manuals on 'Socially Appropriate Technology' whose purpose is to discuss technology from the social aspect. In this they differ from nearly all other appropriate technology literature, which describes materials, techniques, and principles of technology, but not the practice of technology in specific social circumstances, and rarely the organization of specific technical projects.

The subject of this particular manual is the basic technology of horticulture and vegetable-growing, as it applies mainly to family gardens. It deals largely with the way this kind of vegetable cultivation has been encouraged through extension activities initiated mostly by health service workers and nutritionists. In some ways it is almost as if medical practice has expanded to include the practice of a branch of agriculture.

Discussion of technology from this angle depends very much on field reports and case histories which illustrate the connection between administrative, social and technical factors. The case histories used in compiling this manual are indicated in the text using references based on Oxfam file numbers, to which the following is the key.

Reference key

BD 20	Bangladesh, People's Health Centre, Savar, near Dacca.
BD 34	Bangladesh, Seba Sangstha: Rangpur Medical College students' dispensary.
BD 70	Bangladesh, agricultural re-settlement, Haluaghat, Mymensingh.
BRZ 111	Brazil, Teresino, Piaui, experimental garden.
BRZ 166	Brazil, Parnaiba, Piaui, communal garden, housing.
BRZ 165	Brazil, Mossoro, Rio Grande do Norte, community education.
BUR 2	Burundi, Buye Hospital, Nurses' Training Centre.
BUR 16	Burundi, Musenyi Health Centre.
BUR 22	Burundi, Centre d'Entreaide et de Developpement, Bujumbura.
GUA 1	Guatemala, Chimaltenango, integrated development project with World Neighbors.
IS 80	India, NRU at Campbell Hospital, Jammalamadugu.
KN 13	India, Bangalore, Home for the Aged, vegetable garden.
MAL 15	Malawi, NRU at St Martin's Hospital, Malindi
PK 42	Pakistan, Catholic Medical Project, Zairat, Baluchistan.
RHO 34	Rhodesia, Silveira Hospital, nutrition project.
RHO 45	Rhodesia, NRU at St Thereza's Hospital, Chilimanzi TTL.
RSA 3	South Africa, Zisizeni Health and Welfare Association, Kwazulu.
RSA 6	South Africa, Holy Cross Hospital, East Pondoland.
RSA 7	South Africa, St. Lucy's Hospital, Tsolo.
RSA 16	South Africa, The Valley Trust, Botha's Hill, Natal.
UGA 12	Uganda, Mwanamugimu NRU, Mulago Hospital, Kampala.
UP 15	India, Agrindus agricultural project, Uttar Pradesh.
YEM 9	Yemen, MCH Centre, Sana'a.
ZAI 32	Zaire, Vanga Baptist Hospital, health education programme.
ZAI 67	Zaire, CEMEKI Agricultural Programme, Ngidinga.
ZAI 84	Zaire, Kananga, soya bean development.
ZAI 70	Zaire, Oxfam Advisory Team, Kikwit.
ZAM 22	Zambia, Chikuni Nutrition Centre, Gwembe District.

Summary

Conventional agricultural services mainly encourage commercial crop production without much thought for the nutritional consequences of the policies they advocate. Medical services and other organizations concerned with the high incidence of malnutrition in some countries have experimented with a different kind of agricultural programme aimed at helping those who grow vegetables for immediate use by their own families.

The first three sections of this manual describe programmes which aid women gardeners and others who produce food directly for their families. Section 4 discusses the illnesses associated with poor nutrition, and the vegetables most relevant to their prevention. Sections 5, 6 and 7 are concerned in greater detail with vegetable crops and the practicalities of producing them in small gardens.

1. Agriculture, Malnutrition, and the Voluntary Agencies

Nutrition-oriented agriculture

Agricultural development is often planned primarily as part of economic development, with extension and back-up services designed mainly to help the community to earn a larger money income, and to direct more agricultural produce into the national market economy.

This approach is, of course, often beneficial in bringing prosperity to a rural community and in reducing poverty, but may not be of much help in combating malnutrition. This is because economically-oriented agricultural services tend to concentrate on a limited range of crops, often encouraging the farmer to use his land for whatever crops may be most saleable, whether they be cotton, tea, tobacco or food crops (including the staple food). Meanwhile, the wide variety of foods used in many traditional diets tends to be lost as agricultural specialisation is encouraged and as wild vegetables and berries disappear with the clearing of forest and bush. So symptoms of nutritional deficiency may become more prevalent even as money incomes slowly begin to rise.

Development in agriculture is necessary if hunger and poverty are to be eliminated, but the emphasis on such development needs to be counterbalanced to some extent by *nutrition-oriented development,* aimed at broadening the range of crops grown for the family's own consumption. In the hungry season when traditional staple foods are in short supply fresh vegetables can be cultivated so that the whole diet is better balanced. The movement will often be quite opposite to the trend towards specialisation noticeable in most economically-oriented agriculture.

In many communities, this contrasting, nutrition-oriented approach will often appeal strongly to women, who select, cook, and often grow most of the food their families eat. Men, in contrast, often see themselves as earning the family's income rather than providing its food. When this is the case, there is need for a nutrition-oriented agricultural service aimed at women rather than men.

A comment repeatedly quoted is that, although in Africa women may produce 80 per cent of the food eaten at home, they are denied access to agricultural extension services, while at the same time, "they are told by well-meaning nutritionists what food to give their children, without always being told how to grow it; Western advisers assume that all farmers are men".

One aim of this manual is to go some way towards redressing this bias by describing alternative kinds of agricultural extension work which are likely to meet the needs of women rather than men, and which are related to forms of gardening and food production with which women are most concerned.

Voluntary Agencies

Nutrition-oriented agricultural development is not only necessary as a counterbalance to the existing emphasis in agricultural extension; it also provides special opportunities for the voluntary agencies.

In many countries, the development of commercial agriculture is fully provided for, in theory at least, by government agricultural services, and there are few opportunities for voluntary agencies to play a part. So despite the emphasis on agriculture which Oxfam, for example, has encouraged in its overseas programmes, its support for agriculture has failed to expand in the manner hoped for. But on examining alternative kinds of agricultural service, one finds that nutrition-oriented agricultural work, directed to women, is rarely tackled by government agencies, even through nutrition education programmes which some health services may initiate, unless the country has an integrated nutrition programme.

Another reason why voluntary agencies may be able to play a particularly useful role is that this approach to agriculture can be conveniently combined with primary health care and health education work in which many such agencies are already actively involved – indeed, the multi-disciplinary approach required, embracing health, nutrition, and agriculture, is often more easily tackled by a small agency than by a government department which has to cover a whole country.

It would be wrong, of course, to make too sharp a distinction between nutrition-oriented and economically-oriented agriculture. Although in some communities, the women respond directly to nutritional arguments in favour of vegetable gardening, in other communities, and particularly where men are concerned, it is difficult to introduce vegetable growing for purely nutritional reasons. It is often necessary to 'sell' the idea as an income-earning activity with the hope that some non-marketable production will be eaten by the grower.

Also, it is no bad thing if a family can run a successful smallholding which produces a surplus of food for sale *locally.* That in itself contributes to the nutrition of the local community. Some nutrition projects actively encourage this by running market stalls on their premises where local growers can sell their produce; in one instance, the stall is set up at the entrance to a hospital clinic (RSA 6). Another nutrition project buys vegetables from local producers in order to dry them and re-sell the dried vegetables during seasons when fresh produce is unobtainable (ZAM 22). This strategy has had considerable success in stimulating vegetable production in the locality, by increasing the market demand for such produce.

Although it may include economic activity of this kind, nutrition-oriented agriculture differs from commercial agriculture in a number of ways:

a. in growing crops because of their nutritional value rather than because of their market value.
b. in concentrating on gardens of a size which most families can cultivate.
c. in appealing primarily to those who produce the family's food – in many communities, the women.
d. in linking agricultural extension work to health education, social education and community development.

The extent to which this kind of programme aims at women solely or mainly will depend very much on local circumstances. Women are often more highly motivated than men with regard to family needs, and may be more receptive to ideas about 'development' as a result. On the other hand, restricted opportunities for education and travel in search of work sometimes leave women with a more conservative view of development. In some communities, women lead enclosed lives within the household, but in most of the developing countries, women play a significant part in agriculture. Distinctions must be drawn, however, between communities where the women's role is simply to work in the fields as labourers (as often in Asia), and communities where the women are farmers in their own right, making basic decisions about what crops to grow and when to plant them. This latter situation is widespread in Africa.

Field reports from African countries stress the importance of concentrating on women in efforts to improve crop production. One such report comments that "the majority of non-government organizations involved in improving crop and vegetable production launch into such activity by building a centre and training men in agriculture . . . The results of such an approach have on the whole been a complete failure" (ZAI 70). The rather different kind of agricultural extension work needed to help women gardeners can often most effectively be linked to health education stressing the importance of nutrition, and this approach is described in the next section of this manual.

In order to show the different approaches to nutrition-oriented agriculture which are possible where the role of women in agriculture is less central than in Africa, a number of Asian and Latin American projects are also described. In Brazil, for example, the typical approach is through community development work involving both men and women, rather than through health education aimed at women, and this approach is described in section 3.

2. Approaches Based on Health Services

Health education to counter malnutrition

The type of agricultural work to be described in this section has most often originated from the experience of doctors or other health workers who have found themselves treating a great deal of illness arising from poor nutrition. Parallel experiences can be quoted from Africa, the Middle East, Latin America, Pakistan and Bangladesh, where a doctor or nurse has felt frustrated at having to treat only the end-result of malnutrition and has decided to take a much more direct form of action by trying to tackle the deficiencies in availability of food which seem to be causing it. It is in this way that health workers have often become involved in schemes for agricultural improvement or small-scale vegetable growing. In one instance, a doctor caused a school for agricultural extensionists to be set up behind his clinic (GUA 1); in another, soya bean cultivation was introduced with the idea of combating protein deficiencies (ZAI 84);* in a third example, Oxfam gave a spade, a hoe and a wheel-barrow, along with medical supplies, to a mother-and-child-clinic in Yemen which has a demonstration garden (YEM 9).

Not all these projects have used the same methods, and not all have been equally successful – introduction of soya, for example, has not always secured the benefits hoped for. However, there is a pattern of significant if modest benefits to be seen

---

*The two projects referred to here are also discussed in *Growing Out Of Poverty*, ed. Elizabeth Stamp, Oxford University Press, 1977.

in projects which have worked mainly with the mothers of young children in those areas of the world where women are traditionally responsible for growing the family's food crops.

Usually, agricultural work with mothers has been the result of an experience which has shown that nutrition education by itself has had little impact. Even when the connection between nutrition and health is explained in detail, the common experience is that the children do not get better fed, and their health does not markedly improve. Sometimes this is because the mothers cannot obtain the nutritious foods being advocated. Nearly always, though, the mothers need practical help in growing or preparing food, not just the knowledge that a particular kind of food is specially valuable in feeding their children. So in many places, health education activities have evolved to include food production as well as the preparation of nutritious meals, and practical demonstrations in addition to talks given during clinic sessions. Often there are demonstration gardens laid out beside clinic buildings, market stalls at the clinics where people can buy vegetables, and stalls selling vegetable seeds for mothers to sow on their own land.

Even these measures, though, may be insufficient to secure the necessary improvements in nutrition, because what many families need is the opportunity to discuss their own particular gardening or dietary problems on the spot. Sometimes, too, they need material help with fencing to keep animals off their garden, or in reclaiming a neglected plot. Thus some health centres and hospitals have gone far beyond running a demonstration garden, and have developed a follow-up or extension service whereby mothers of families with nutrition problems are visited in their homes by health workers expert in nutrition and gardening – or mothers who already have a garden are visited by an extension worker who can suggest improvements or help solve particular problems. Some health centres have even organized revolving loan schemes which help people pay for fencing or the initial land preparation involved in first setting up a garden.

Types of health service gardening projects

Once a health service organization has become involved in practical agricultural or gardening work, the way this is arranged depends very much on the type of health service involved. A hospital-based organization will obviously adopt a different approach from one depending mainly on village-level health workers, while rural clinics and health centres occupy an intermediate position. Nutrition rehabilitation units or centres (NRUs) where mothers with malnourished children may stay with their children while nursing them back to health have a very specific need for productive gardens as an integral part of their operation.

Some examples of these different kinds of health service organization are presented in table 1. It will be noticed that in all cases, vegetable growing is one of a number of practical activities linked to education work aimed at mothers. Where the projects differ most noticeably is in the degree to which they support extension work and foster activities at village level. Although such work, influencing individual mothers in their homes, is the most difficult aspect to organize, it is likely to be the most effective.

Table 1 also indicates what agricultural staff are employed by the health services

Table 1: Agricultural Activities of some Health Service Organizations

1. Type of Organization						
(a) Nutrition Rehabilitation Units or 'Villages' (NRUs) attached to Mission Hospitals		*(b) People's Health Centre;* hospital & rural clinics with co-operative health insurance scheme	*(c) Outlying Health Centres or Clinics in rural areas*			*(d) Training schools for Health Workers*
South Africa (RSA 6, RSA 7)	**Malawi (MAL 15)**	**Bangladesh (BD 20)**	**Burundi (BUR 16)**	**Pakistan (PK 42)**	**South Africa (RSA 16)**	**Burundi (BUR 2, BUR 22)**
2. Educational Activities						
mothers taught veg. growing with cooking, nutrition.	mothers taught veg. growing with cooking, nutrition.	some mention of agriculture in general health education.	mothers at clinic told about protein-rich vegetables.	samples of salad veg. and charts showing what veg. to eat.	extensive demonstrations of cooking, gardening for clinic patients.	nurses taught a little gardening in 2nd year public health course. Village health workers also taught gardening.
3. Practical Activities at Centre						
mothers in NRUs do gardening, produce food for own children. Stall at hospital entrance (RSA 6) sells seeds and vegetables.	mothers do gardening, also paid labourer.	hospital has field growing soya beans; soya flour products used in canteen. Stall sells soya milk, soya biscuits.	health centre has garden growing soya and green veg. Soya milk available.	demonstration garden.	demonstration gardens at several places. Stall at health centre sells garden produce, seeds.	the trainees do some work in demonstration gardens at the training centres.
4. Activities in the Villages						
help is offered to villagers with fencing gardens, but both projects are criticised for lack of extension work.	none yet	'para-agros' to work alongside para-medicals and sit on Village Health Committees.	soya seeds given away; vague encouragement with soya and groundnut cultivation.	none yet.	good, effective extension work; loans for fences; practical help with fencing, deep trench cultivation and fish ponds.	intention of having demonstration gardens at all rural health centres throughout country.
5. Agricultural Staff Employed						
RSA 7: 1 agricultural demonstrator + 2 gardeners + 1 part-time	labourer (no details)	1 agricultural labourer + 2 para-agros; medical staff also do gardening.	nurses do gardening.	nurse and priest do gardening.	1 agricultural demonstrator + 2 or 3 assistants, + part-time schoolboys.	training given by government agriculturists, health workers do gardening with help from women attending clinics.

represented. Where an agriculturist is employed at a hospital, health centre, or nutrition rehabilitation unit, he or she looks after the demonstration garden there, and is typically known as an 'agricultural demonstrator'. Where extension work is carried out in the villages, the demonstrator will also be responsible for this, visiting mothers in their homes in order to emphasize points put over during health education sessions at the clinic, or to discuss practical problems in starting or improving a garden at the family home. Other staff employed may be gardeners or labourers who help maintain demonstration gardens, or who may help mothers clear the ground when starting a new family garden, but who do not have any responsibility for teaching or extension work.

In some projects, medical staff help in the garden, usually because funds to pay the wages of gardeners are lacking, though sometimes as a matter of policy, in order to keep doctors and nurses in touch with the non-medical side of their organization's work. In hospitals or other institutions where handicapped or disabled or old people find shelter as long-stay inmates, they may contribute some part-time work in the garden (as with RSA 7, KN 13), and older children may also help during school holidays (RSA 16).

Perhaps the most typical kind of health service organization to become involved in agriculture is the outlying health centre in a rural area, of which three are represented in table 1. One of these (RSA 16) is the oldest gardening/nutrition programme which Oxfam has supported, and it has stimulated many other health services, mainly in the southern half of Africa, to follow its example.

The agricultural activities of this health centre began in 1959, in a rural community where most families kept cattle and grew maize and a few vegetables on plots of land near their homes, but where almost nobody had a fenced garden with adequate soil conservation measures, or used compost to maintain soil fertility. By providing gardening demonstrations at the health centre, backed by extension service visits to mothers in their homes; and by supplying posts and wire for fencing, and offering loans to help pay for them, this organization has had a very marked impact on the agricultural and nutritional practices in the area it serves.

As evidence of this, in the 15 years from 1959 to 1974, some 240 families received help from the health centre in creating fenced gardens, and it seems that another 400 gardeners have made use of the health centre's methods without receiving any direct assistance. Apparently as a result of this programme, the health centre recorded a marked decrease in the incidence of malnutrition, and in diseases in which poor nutrition reduces resistance, such as tuberculosis. Kwashiorkor cases seen at the clinic fell from 145 in 1963 to only 5 in 1973. The incidence of kwashiorkor remained high in neighbouring areas not served by the health centre, and so the decrease can be attributed mainly to the greatly increased cultivation and consumption of vegetables.

A somewhat different approach may be observed in places where para-medical workers are employed, and live in the villages they serve. Where they are primarily concerned with nutrition, these health workers are now often known as 'nutrition scouts'. Alternatively, where para-medical workers are complemented by village-based agricultural advisers, the term 'para-agro' has been coined (BD 20). Yet another name used for village health workers who deal with agriculture coupled with nutrition has been to call them 'agronuts' (ZAI 70), or agro-nutritionists. Both men and women have been employed in these roles.

Evolving approaches to malnutrition – NRUs and nutrition scouts

It will be evident from this that health services have generally become involved in agriculture by a steady progression of developments, beginning with nutrition education, then enlarging its scope by starting demonstration gardens, and finally organizing extension work and follow-up activities reaching mothers in their homes. A further evolution in recent years has led to the development of two more sophisticated approaches to the prevention of malnutrition: nutrition rehabilitation units and nutrition scouts. In both approaches, there is an important role for stressing the production of more food from the family garden.

Nutrition rehabilitation centres, villages, or units – NRUs for short – represent a specialist response to malnutrition in which the agricultural component has often been particularly stressed. The concept of the NRU originated in the late 1960s, after it was realised that many children given curative treatment in hospital for malnutritional disorders (such as kwashiorkor) suffered a recurrence of the condition within a few months. It seemed that when children were taken into hospital, mothers gained the impression that it was some specialized 'treatment' which cured their children, and did not fully appreciate that the main factor in the cure was improved feeding – i.e. more food in total quantity and a better balanced diet. Consequently, when children returned home, the mothers did not recognise or understand the need to maintain the improved diet, so often the malnutrition symptons reappeared with less hope of cure. Surveys of children treated in hospital have helped to reveal the extent of the problem.

For example, limited surveys in Iran, India and the Philippines indicated that between a third and a half of the malnourished children 'successfully treated' died of malnutrition within six months of discharge from hospital. However, pioneer work in Venezuela and Uganda (UGA 12) showed that if the mothers could be housed with their malnourished children in a unit outside the hospital, and could be made responsible for feeding the children back to health under supervision, not only would the children recover, but the mothers would learn what kind of food was needed and how to prepare it.

Where this approach is used, mothers and children are housed in village conditions, and use equipment similar to what is available at home. Thus, the training received by the mothers is immediately applicable to home conditions.

NRUs generally have a vegetable garden and often a poultry unit as well, and the mothers are given practical training in growing, preparing and cooking vegetables, and in using eggs and poultry meat. In one NRU, the women are expected to do two hours' work in the garden each day; they are taught how to sow and transplant, and how to make compost, and later, they use vegetables from the garden for making meals to feed to their children (RHO 45).

In the part of southern Africa where this NRU is situated, most families have, or would like to have, a small field of up to one hectare in extent planted with maize, beans, pumpkins and a few vegetables. Surveys in some communities show that the vegetable production is being more systematically pursued as a result of teaching done at NRUs. Thus, more vitamin-rich leafy vegetables are being produced to supplement staple foods, and some families who do not have a field of their own have at least been able to start a small garden. In one locality, out of 60 mothers who had been through NRUs, eight who had previously been growing vegetables tried out new crops which they had seen at the NRU and 13 who had not grown vegetables before started gardens of their own.

Experience has shown, however, that if mothers are to transfer the lessons learned to their own land at home, they need help with seeds, plants, fencing materials, and sometimes tools. Visits from an extension worker to back up this help are also necessary. Sometimes the materials are sold to mothers, with loans available to help where necessary (RSA 7); sometimes, material help is given free, on the understanding that the mother will in due course make a gift of produce from her garden to the NRU. One NRU gives each discharged mother two kinds of seeds – one root vegetable and one green vegetable. Any mother who then successfully starts a garden of her own is later given a guava tree to plant in it (RSA 6).

Agricultural activities associated with NRUs seem to have met with modest but significant success in Africa, particularly where the traditional role of women includes the growing of crops for family consumption, and especially where there is active follow-up of mothers by extension workers. In India, where women have less responsibility for crop production, NRUs pay less attention to gardening, though maintaining a practical emphasis on teaching food preparation and cooking.

As to the success of NRUs in preventing the recurrence of malnutrition, one Indian survey (IS 80) has shown that less than 10 per cent of children who had passed through the NRU died within a year. However, experience elsewhere has not been so good, with surveys at two NRUs in Africa revealing a record little better than that obtained with curative treatment. The number of children dying within a year of leaving these units was found to average 28 per cent, with most deaths arising from illness associated with poor diet. There are several reasons for this disappointing performance. One is that, by the time they were seen in the hospital clinics, many of the children were too malnourished to recover simply by giving them good food. They needed medication, and the NRUs had to provide curative treatment almost comparable to a children's ward – the ideal of separating them from the hospital was thus not achieved.

Another problem is that taking mothers with their babies away from home for three weeks can create difficulties for the rest of the family, and can cause extra worries for the mothers themselves (a number of whom walked out of NRUs without completing the course). Mothers who are worried about what may be happening at home, even if they do last the course, will not concentrate on what is being taught, and may therefore not learn very much.

As a result there is now a fairly widespread opinion that the NRU approach has not lived up to the hopes of those who pioneered it. Teaching mothers that good food is the cure for malnutrition can perhaps only be really effective if this teaching is carried out by health workers who visit the mothers in their own homes, where they can detect the symptoms of child malnutrition at an early stage, before medication is needed; and where advice about diet and cooking can be based on what the mother actually has available.

Efforts have been made in some countries to train village health workers to do this. These are the "nutrition scouts" previously referred to. After a decade of experience with NRUs, it is now generally felt that they deal with too small a minority of mothers, and are not even very successful with these. What is more urgently needed is a preventive service for the generality of children in rural areas, nearly all of whom may come near to suffering from malnutrition at some time in their first five years. The locally-recruited nutrition scout, who can examine and weigh babies in their homes, and who knows the mothers as neighbours and not only in the context of health care, should be able to offer the 'grass-roots' level preventive service which seems to be needed. Considerable ingenuity has gone into devising simple, basic equipment for nutrition scouts to use, comprising scales for weighing children, record cards with growth curves printed on them, vitamin A capsules, iron capsules for anaemic mothers, and materials for mixing a sugary, salty drink for rehydrating children with diarrhoea. Much of the pioneering work has been done by UNICEF, notably in East Africa. A nutrition scout in one of their programmes is simply a villager who has been shown how to recognise the early signs of malnutrition, and how to treat children with diarrhoea and other common diseases. She (or he) visits mothers and children in her own village, recording babies' weights (or arm circumference measurements), encouraging attendance at the local clinic, talking about hygiene and cooking, and the possibility of growing more vegetables. Most important, the nutrition scout must set a good example in the village by making sure that her own children are well, and that her vegetables and poultry are flourishing. Scouts may also make follow-up visits, or do extension work with mothers identified at clinics as having specific problems with feeding their children.

Experience with nutrition scouts is still limited (UNICEF, RHO 34, RHO 45), and opinions differ as to whether women or (married) men are more effective in this role. Men who are widely respected in their own communities have been very effective, but are at an obvious disadantage in discussing some medical or family planning matters with mothers. Probably the scouts should be chosen locally by village meetings, rather than being picked out by hospital authorities, so that the question of whether a man or woman gets the job is left for the village to decide.

The suggestion has also been made that where NRUs already exist, and where it is desired to move from the NRU concept to the nutrition scout approach, the buildings and gardens formerly used as an NRU could very appropriately become a training centre for nutrition scouts, and a centre from which villagers could obtain seeds, fruit tree plants, and fencing wire at low cost, the materials having been bought in as a bulk purchase.

3. Approaches Involving Community Participation

Village committees and community associations

The two gardening and nutrition programmes in Asia included in table 1 (BD 20 and PK 42) are significantly different from the African programmes so far emphasized, in that neither of them is aimed exclusively at mothers, but at a broader cross-section of the community. Of the two, the Pakistan project retains the closest similarity with the African programmes, being based mainly on clinic sessions, and stressing the nutrition of children. One may compare it with a group of clinics in Yemen run by a mother and child health centre. The centre has three main departments: ante-natal, sick children and nutrition. The nutrition department is chiefly concerned with educational activities, and like the African projects, includes food preparation, cooking and gardening in a programme aimed at mothers.

In contrast, the programme in Bangladesh which has developed an agricultural dimension (BD 20) includes a health insurance scheme run on co-operative lines with village health committees to co-ordinate local activities. In this way, the whole community becomes involved in the health programme. Six of the village committees each include one committee member responsible for agricultural activities. The aim is for these committee members to be given some training in agriculture and extension methods to make them more effective as "para-agros", or village-level extension workers. The village committees are a recent development in an ambitious, multifaceted programme and their role is not yet fully evolved. But village health committees or development committees do seem to provide an alternative way of making the link between nutrition and agriculture, since they take an integrated view of the development of a single community. This, then, is an approach which may be effective in places where women do not have any particular responsibility for food crops, and where the activities of mothers do not provide a natural focus for making the nutrition-agriculture link. However, even where there are opportunities for gardening programmes oriented towards mothers, village committees may have a role to play in broadening the emphasis of health work, and in creating opportunities for integrated forms of development. By organizing health projects for their members, committees may be valuable as agents for self-education.

There is also a useful contrast to be drawn between the activities of a village health committee and the traditional kind of medical programme. The latter is conventionally operated on a "provided service" basis, with local people presenting themselves at a clinic, or being visited in their homes, but not being expected to participate in running the programme, nor to voice opinions about the kind of service which should be offered.

But with the growing emphasis on preventive health work, medical staff are slowly learning that there is a real need to involve the community more fully. When it comes to recruiting nutrition scouts, getting latrines dug, or finding volunteers to help with the local clinic, it becomes essential to listen to local views, and to introduce a degree of community development into the programme. Some health projects have evolved in this direction without conscious planning, but others (e.g. RSA 3) have quite deliberately attempted to move from a "provided service" approach to one of community participation and social development.

Agricultural aspects of health programmes have also usually been offered as a "provided service", although of course, they demand a very active response from the individual gardeners or smallholding cultivators. However, where community development activities lead to the formation of a village development committee or health committee, the committee itself can organize talks and demonstrations on gardening, and can perhaps arrange for people to help each other with heavy jobs such as fencing.

Where there are nutrition scouts or other para-medical or para-agricultural workers living in the village, they usually become members of village committees, and are able to awaken interest in food production projects. One instance of this (ZAI 67) is where agricultural extension workers are themselves farming in two of the villages. They are members of their village committees, and a visitor was "favourably struck by the influence of the village development committee on the villagers, and their awareness of the relationship between health, nutrition and agriculture . . . "

The importance of involving the people in these programmes is illustrated by one programme which has undergone the transition from "provided services" to a community development approach (RSA 3) – a clinic run by a church organization was experiencing problems with an orchard attached to it, and trees were beginning to die off. Technical advice was obtained about the care of the trees, but at first it proved difficult to obtain anybody's co-operation in implementing this advice. People coming to the clinic did not seem willing to help, because they had never participated in running the orchard previously, and regarded it as a church project and therefore not their responsibility. However, when the church handed it over to the community and an agricultural committee was formed to run the orchard as a community enterprise, attitudes began to change because there was now a more positive means through which local people could take an interest in the orchard.

Community associations in a semi-urban environment

A somewhat different version of this more broadly-based approach aiming at the whole community is to be found on the fringes of some of the towns of north-east Brazil, notably in the poorer suburbs of Parnaiba (population 60,000) and Mossoro (40,000). Here, the impact of social and health programmes initiated mainly by church organizations has been the development of community movements and co-operative efforts in a wide range of activities, many of them linking together agriculture, nutrition and health.

The way these activities develop is typically that a team of extensionists begins work with a community, offering various adult education and literacy courses, and encouraging practical measures to improve health, nutrition or housing, or to generate income. As the people take up these activities, express their preferences and take the lead in some of the work, a community group develops which may ultimately be able to stand on its feet as an autonomous community association.

One team of 30 extensionists covering several depressed suburban areas is involved in an impressive range of activities, including fruit growing, pig raising, nutrition education, communal vegetable growing, crop storage buildings, latrine construction, and handicraft production (BRZ 165). It is up to each community to decide which of the many courses and activities on offer are most suited to its particular needs, so in different areas, different projects on this general list are being put into effect. Links between agriculture, nutrition and health are emphasized in the process mainly because these are among the most urgent problem areas facing these communities.

Another example of this approach is provided by a community group which has projects in vegetable growing, fruit trees, chicken raising, and latrine construction (BRZ 111). Yet another similar project began with a scheme for people to help each other rebuild their own houses, and followed this up with a plan to fence in the back-yards of the houses and use them for vegetable growing. This gardening work was also to include a communal vegetable plot, which would provide the focus for demonstrations of improved gardening methods (BRZ 166).

The particular character of these projects in Brazil arises partly from the fact that they are organized by local agencies staffed by Brazilians, and not by expatriate medical workers who see problems in terms of their own special subjects. It is also more appropriate for people from the country concerned to see practical measures in terms of broad social objectives, and interpret them in that light.

Another factor, though, is that outside the areas where women have a traditional responsibility for food production – mainly in Africa – the link between vegetable growing and health is not likely to be made successfully just by approaching mothers attending clinics. A wider sector of the community has to be involved. And in an area where many of the poor have neither full-time employment nor sufficient land for farming, growing vegetables on small plots of unused land may be coupled with odd jobs and handicraft production as a means of securing subsistence for the family. Indeed, with a large urban market for food nearby, growing vegetables may be seen as an income-generating activity, with only a small proportion of produce actually being consumed by the family.

Community gardening

The African projects for promoting vegetable growing previously described have been based on two assumptions. One has been that an approach to women in their role as cultivators as well as mothers could be effective. The second assumption has been that all families have access to land, which is widely true in rural Africa. There have sometimes been problems in that many African systems of land tenure are based on concepts of communal ownership, in which the fencing of plots for permanent use by an individual family is not acceptable. However, where it is necessary to fence gardens to keep off animals, experience shows that objections can be overcome if people are given time to discuss the issues, and if customary authorities within the community are consulted.

In the poor suburbs of Brazilian towns, the situation is completely different. People do not usually have rights to land suitable for cultivation. But the plots on which their houses stand do include back-yards which are often big enough for some useful gardening. Indeed, even two or three square metres of good soil, wisely used, can contribute vitamin-rich crops to the family diet which may be of much greater nutritional value than their negligible financial worth.

Putting these back-yards to good use, either in keeping chickens or growing vegetables, is in fact relevant both to the income and the nutrition of these Brazilian families. In two cases, attempts have been made to supplement the potential of the back-yards by setting up communal gardens which householders can use in conjunction with their home plots (BRZ 166, BRZ 111). In one of these already mentioned (BRZ 166), the project grew out of an earlier one in which the people helped each other build new houses for themselves, then fenced back-yard plots and began using them to grow vegetables. The aim of the communal garden was to support this activity by providing a training ground where improved techniques suitable for use in the back-yard plots could be learned and to use them as a source of plants and seedlings for the household plots. Other vegetables were produced which could be marketed in town. Income from such sales was divided, with 40 per cent going to the people who worked in the garden, and 60 per cent being retained to buy seed and fertilizer.

This project has not worked out as planned, although useful quantities of vegetables are being grown. It does, however, illustrate the potential of a communal garden as the focus of a gardening co-operative. Advantages to be gained from this kind of co-operation might include the possibility of sharing expensive packets of seed among owners of several back-yard plots, each of whom might require only a very small quantity.

In instances like this, the existence of a communal group, working together, may be crucial in securing the necessary land. An approach by the community to the municipality or other owners of unused land could have the effect of releasing it for gardening where an approach by an individual would be ineffective. The project just mentioned (BRZ 166) obtained the use of some waste land belonging to the church, and devoted a lot of effort to reclaiming it for their garden. Where school gardens exist, it may be possible to use them in conjunction with back-yard plots to increase overall vegetable consumption and production in the community. In one African project, children were given plants from the school garden for them to grow in the family plot at home.

Communal gardens may sometimes be cultivated as allotments, with each family growing vegetables in its own individual part of the garden, rather than working co-operatively on the same plot. In southern Africa, the allotment system has developed in some places because the high cost of fencing materials has made it too expensive to fence individual household plots; a single community garden in which individuals have their own plots was much more economical (RSA 16). Also, wherever money has to be spent to provide water for irrigation, this could prove to be cheaper for a communal garden than for separate household plots. Communal gardens may be run by a committee formed from the holders of plots to control repairs to fences, water for irrigation, changes of tenant, and any co-operative seed-purchasing arrangements.

In one such scheme, set up by a dispensary/health centre in Bangladesh, there was discussion about running the garden as a "co-operative" with an initial loan to rent the land, buy seed and fertilizer, and erect fences (BD 34). After the garden had begun to produce vegetables, allotment holders would then pay back the loan and gradually take on full responsibility for the rent.

In all such schemes, it is obviously important to ensure that the gardeners have security of tenure, which means that the co-operative or allotment-holders' committee must be properly constituted in the eyes of the law, able to hold land legally on behalf of its members. In some cases, Oxfam has given grants to pay for legal registration of community associations involved in market gardening (BRZ 190). It is significant that social education programmes have also included courses on civil rights and land tenure, and have included practical projects on the demarcation of land boundaries in their work. In some countries, where the poor do not have security of tenure on the land they use, or where they are likely to find it difficult to exercise such rights as they do possess, these activities connected with the legal side of land use may be just as crucial for the establishment of vegetable growing, as far more practical work connected with fencing or cultivation.

4. Choice of Crops for Improved Nutrition

Experiences of malnutrition

The chief characteristic of nutrition-oriented gardening as opposed to economically-oriented agriculture is that the choice of crops to be grown does not depend on their market value, but on an awareness of dietary needs in the community or family concerned.

Surveys of diet-related illness in different parts of the world reflect a great variety of local conditions, but two points are repeatedly noted throughout the developing countries. Firstly, there is the struggle to grow *enough* food, especially the staple food crops (cereals and starchy roots) – a struggle which in years of drought or pest is not always successful. Secondly, there is often a deterioration in the quality of diet in places where commercial pressures encourage an increase in the use of processed foods, as in peri-urban areas where the easy preparation and cooking of processed foods and availability of cash are factors in dietary change, and where lack of suitable land and time for gardening also play a part.

In certain instances, both these problems exist side by side. There, one may paradoxically find the illnesses caused by under-nutrition due to insufficient production of staple foods in the same community as the illnesses associated with increased consumption of some of the common processed foods. For example, health workers in the South African "homelands", such as the predominantly rural Transkei and Kwazulu, are confronted with a very high incidence of tuberculosis worsened (though not caused) by under-nutrition; and many children suffer from kwashiorkor because too little food is provided for them. But there is also a dramatically rising incidence of the diseases usually associated with Western habits of consumption – peptic ulcer, diabetes, appendicitis and coronary thrombosis.

To some extent this pattern of illness reflects the diet eaten by migrant workers while they are away in the towns, but it seems also to affect people living in the rural areas, where a wide range of processed foods is available from trading stores and shops. Less milk, meat and other sources of animal fat are consumed than in the traditional diet, and local medical reports attribute the increasing incidence of "Western-type" diseases, including coronary thrombosis, to the growing consumption of sugar, white flour, white maize meal, and similar processed food (RSA 7).

So the situation may be summarised by saying that a decline in traditional methods of food production has led some people to exist on insufficient food and others to resort to unbalanced diets in which certain processed foods are over-represented. Kwashiorkor is a serious and prevalent result of under-feeding in children, whereas, in contrast, the increased consumption of sugar and refined flour is associated with the appearance of the so-called "diseases of affluence" in communities which are in fact very poor. Diets of this kind are also likely to be responsible for vitamin and protein deficiencies.

Because kwashiorkor and protein-calorie malnutrition generally were formerly attributed to a shortage of protein rather than to insufficient food (which is the most recent view), nutrition-oriented agricultural work in these parts of Africa has tended to stress protein-rich foods, including beans of all kinds, fish (from fish-ponds), poultry and rabbits (RSA 3, RSA 16). An effort has also been made to persuade people to eat more vegetables of all types and to reduce their consumption of processed foods proportionately. Thus medical projects in the South African "homelands" have been responsible for a considerable amount of innovation in the borderland between agriculture and nutrition as a result of the exceptionally serious nutritional problems which exist there, and this high level of innovation is reflected in the number of programmes in South Africa which are quoted in this manual.

A rather different example is provided by statistics collected by a medical worker in the Baluchistan region of Pakistan, where a wide range of illnesses, including anaemia, respiratory infections, and constipation were regarded as the outcome of poor diet. In this project (PK 42) the choice of vegetables for a gardening scheme was motivated mainly by data on the poor health of pregnant and lactating women, associated with deficiencies of iron and calcium in the diet. The high infant mortality rate in the area was partly attributed to the prevalence of anaemia in mothers, and an associated lack of mothers' milk. Thus the gardens started by the project organizers were devoted very largely to spinach, which is notably rich in both calcium and iron – though it should be noted that these minerals are not effectively absorbed by the body unless a balanced diet is being eaten which includes, among other things, adequate fats and vitamin C. Thus over-concentration on spinach could have been self-defeating and, indeed, this particular project did also try other crops (notably tomatoes) and included poultry-raising aimed at egg production. In parts of India where many people show symptoms of iron deficiency, their diet has been found to contain ample supplies of the mineral, and the symptoms have been attributed to an inability to absorb the iron, possibly due to a shortage of vitamin C.

Yet other evidence of poor nutrition is provided by a very detailed World Bank study of a group of road workers in northern India. Some of them came from villages close to where the road was being built, while others, from outside the area, lived in a camp on the site and were supplied with a standard daily ration by the road-building agency. This ration contained no fresh fruit or vegetables, and consequently it was not surprising that around 33 per cent of the men showed symptoms of vitamin A deficiency (including night blindness). There were also deficiencies in vitamins of the B complex. The men living at home had a more varied diet, though barely sufficient in quantity (i.e. in calories) for the heavy work they were doing. Their B-vitamin intake seemed adequate, but many of them also showed signs of vitamin A deficiency.

Most leafy green vegetables, particularly the dark green varieties, contain carotene or pro-vitamin A, as do carrots, sweet peppers, and some tomatoes. The green vegetables also supply vitamin C, and in many cases, iron and calcium as well. In India, nutritionists estimate that if vitamin needs are to be met, existing diets should include an average of 235 grams of vegetables daily for each person, of which 100 grams should be leafy green vegetables. The prevalence of vitamin A deficiencies, widely reported from Oxfam projects, is easily understood from estimates that in India as a whole, average vegetable consumption is only 40 per cent of the above requirement, with consumption of leafy green vegetables reaching only 20 per cent of what is needed.

More detailed studies in Hyderabad show that while urban Indians are able to buy significant quantities of green vegetables, rural Indians, at least in the areas studied, grow hardly any vegetables at all for themselves. The survey may have neglected the wild vegetables which are collected, or the use of green leaves from trees or staple crops, but it did also note that in some quarters, people regarded vegetables as a low-grade or unfashionable food. There is clearly an opportunity here for nutrition education programmes of medical agencies to combat this prejudice.

The estimate that average leafy vegetable consumption in India is less than 20 per cent of what is thought necessary is sufficient in itself to explain the prevalence of vitamin A and vitamin C deficiency diseases. Eye complaints associated with vitamin A deficiency are commonly found; children are badly affected, and many go blind as a result. As a short-term measure, the Government is promoting a programme of periodic doses of vitamin A for children under five; the body is able to store large doses in the liver and use it as required during a period of about 6 months. However, a much more satisfactory solution, conferring other health benefits as well, would be far more widespread cultivation and consumption of vegetables. People in many other developing countries, including several in Africa, experience similar vitamin A deficiencies, though rarely on the scale seen in India.

Choice of vegetables for garden projects

Examples of the kind cited in the previous section, coupled with the attention often given to kwashiorkor and vitamin deficiencies in young children, has led to two main emphases in the choice of vegetables for garden projects:

Firstly, on green vegetables, carrots, sweet peppers, tomatoes and brinjals (i.e. egg-plants or aubergines), all of which can contribute calcium, iron, pro-vitamin A, and vitamin C to the diet.

Secondly, on developing new sources of protein, including beans and other legumes, especially soya, but also extending to the use of garden plots for poultry and small animal projects designed to produce meat and eggs.

Although recent thinking in nutrition confirms that very great benefits can stem from small domestic vegetable gardens, there is a growing criticism of projects which concentrate on only one item in the diet which seems to be deficient, whether that is protein or one of the vitamins. It is being recognised more and more that the most beneficial diet is likely to be the most varied one, drawing protein, minerals and vitamins from as wide a range of sources as possible. Concentration on just one rich source of protein (e.g. soya beans) or of minerals (e.g. spinach) may produce food which the body simply cannot use efficiently. Diet needs to be balanced, not only in terms of food values, but also in terms of constituents which aid digestion and absorption of food. In the past, conventional nutritional theory has emphasized food values (in terms of calories, and weights of protein, minerals and vitamins present), but has neglected some of these other factors. The result has sometimes been that nutritional diseases have become more common in places where nutritionists have had most influence.

In many developing countries, the traditional diet at its best has developed the necessary balance and variety within the constraints of available resources, and it has done this more successfully than can be done at present by scientifically balancing one foodstuff against another. Thus one approach to the choice of vegetables for a garden project might be to study in detail what food is eaten by families whose good health shows them to be well-nourished, and by those, often of the older generation, who have adhered most firmly to the traditional food culture and crop production patterns. When such a survey is made, attention should be paid to wild vegetables included in the diet as well as cultivated ones, and to seasonings used in small quantities which may be providing some vital nutrient. With regard to beans, it should be noted whether they are harvested as green, immature vegetables or when mature and dry; whether pods and leaves are ever eaten, and whether bean sprouts are produced. Note should also be taken of whether people eat the leaves of root vegetables, or of melons and pumpkins, or even whether the leaves of trees are cooked and eaten. Customs followed with regard to drying and storing fruit and vegetables for use out of season are also relevant.

Although in many cultures, conventional green vegetables such as cabbage or spinach are not grown at all, the leaves of other crops, coupled with the use of wild vegetables, provide an ample leafy content to the diet. The leaves of sweet potatoes, melons, cassava, and some beans (e.g. *Phaseolus vulgaris*) are all used as vegetables in parts of the world. In some areas in West Africa, leaves of the baobab tree, and in the Philippines, leaves of papaya and cashew nut trees are cooked and eaten as green vegetables, providing considerable amounts of protein.

Because of these practices, the absence of conventional green vegetables from traditional agriculture need not imply a dietary deficiency. If there is widespread malnutrition in the community, it may be that some of these traditional foods are being neglected owing to the pressures of "modernization", or it may be that some leaves, roots and berries are no longer available because deforestation has destroyed the habitat of wild vegetables and has removed trees whose foliage was a source of food. Reports from an Oxfam project in India document one locality where wild vegetables have become scarce for these reasons (UP 15).

In such circumstances, it may be that the best role that a gardening programme can play is to encourage people to value their traditional vegetables, to strengthen neglected traditions, and to cultivate vegetables which are disappearing in the wild state. A survey of existing practice and custom will be the best way of deciding the details of any such policy for a particular area. Another purpose which a gardening project should try and fulfil is to counter the trend to specialisation which tends to dominate economically-oriented agriculture, and instead stress the importance of variety in people's diets – even a two-metre-square back-yard plot can contribute some variety to the family's food, even if it cannot add many calories.

In the light of these considerations, the main categories of plants which should be considered in planning a garden programme can be listed as follows:

1. vegetables traditional in the area but not sufficiently emphasized in the commercial sector, e.g., okra (in parts of Africa), brinjals (in India, with local species and varieties in other countries);
2. wild vegetables traditionally eaten but no longer plentiful, e.g. several of the amaranths in West Africa (see the detailed notes on crops below);
3. trees with edible leaves and fruit may usefully be planted in odd corners or places where they will not overshadow vegetable plots, thereby helping to counter the effects of deforestation;
4. vegetables which are not traditional to the area should be introduced cautiously where there seems likely to be a particular benefit from their use, e.g. carrots in areas with a vitamin A shortage, or suitable kinds of spinach where there are deficiencies of iron and vitamins A and C.

In addition, tomatoes and sweet peppers are widely grown, and are being absorbed into many food cultures to which they were originally foreign – indeed, the tomato is perhaps the world's most popular vegetable. Both crops should be encouraged as sources of vitamins and minerals.

Reviewing the vegetables actually grown by nutrition-oriented gardening projects, including the ones described in sections 2 and 3 above, it would appear that nearly all these projects have concentrated on "introduced" or "foreign" vegetables, such as cabbage, carrots, Swiss chard, spinach beet, and soya beans; scarcely any has placed significant emphasis on the traditional vegetables of the locality. One long-established project (RSA 16), after years of concentrating on Swiss chard and protein production, has encouraged a survey based on interviews with the older women in the community, seeking their knowledge of local wild vegetables and traditional recipes. This is the kind of activity which is now seen as necessary at the start of a project. The danger of introducing new types of vegetable without such background enquiries is, firstly, that it may lead to the use of foods in combinations which the body cannot use efficiently, and, secondly, it may encourage people to regard introduced vegetables as the most desirable kinds and so to neglect their own food resources. In general, the best policy with vegetables which are new to an area is to introduce one species at a time alongside traditional crops. Some effort should then be made to see whether the introduced crop does add usefully to the variety of foods eaten, or whether expected health benefits really do emerge, difficult though such evaluations may be.

Staple food crops, carbohydrates and protein

The discussion so far has been confined to vegetables in the narrow sense, such as tomatoes, carrots, and the leafy green crops. Little has been said about staple food crops, which are the main source of carbohydrates and of energy (i.e. calories) in most diets. Neither has much been said about beans and other legumes, which often form a considerable part of traditional diets and are widely valued as a source of protein by nutritionists.

The provision in quantity of staple carbohydrate foods will not usually be an important function of the gardening activities fostered by nutrition projects, because in most districts, staple food crops are already grown on a large scale, or can be readily bought by those without fields of their own.

However, because of the importance of variety in the diet, nutrition-oriented agriculture or gardening can usefully aim to diversify the sources of carbohydrate and protein available. So in an area where the staple food is a cereal, sweet potatoes or other tubers and roots might be included in gardens; and where the staple food is a root crop, maize might be grown in the gardens.

This policy has some insurance value in seasons when the staple crop does badly if the alternative carbohydrate crop fares better. But particularly where tubers and roots are the staple, the growing of a little maize or sweetcorn can improve the protein content of the diet, which may otherwise be low. It has been said that, "in a balanced diet, protein should be supplied from as many different sources as possible", particularly if one's main source of protein is vegetable rather than animal.

Almost all vegetables, roots, and cereals contain significant quantities of protein. Cassava root and bananas are the only staple foods whose protein content is negligible, and it is people who depend heavily on these crops who most frequently suffer serious protein deficiencies. Green vegetables, in particular, are often very good sources of protein ("leaf protein"), though this is not always recognised, mainly because tables of food values express the protein content of greens on a "fresh weight basis". Comparing the amount of protein in food after cooking, however, one finds that many green vegetables have as much protein as most beans (table 2). For example, cassava leaves contain as much protein as boiled French beans (*Phaseolus vulgaris*); broccoli and the edible pods of cowpeas are on a par when cooked.

Table 2 Percentage by weight of protein in vegetables when cooked and ready to eat*

(where botanical names are not quoted, they may be obtained from table 3)

High protein content	
Soya beans, boiled	12.8 per cent
Mung beans, boiled	11.0
Papaya leaves	8.0
Lima beans, boiled	7.4
Cassava leaves, boiled	7.2
French beans, boiled	7.2
Moderate protein content	
Spineless amaranth	4.4 per cent
Broccoli (*Brassica oleracea*)	3.5
New Zealand spinach (*Tetragonia expansa*)	3.3
Cow peas, edible pods, boiled	3.0
Sweet potato, boiled	2.6
Rice, boiled	2.2

*Simplified version of a table given by H.D. Tindall in *Food Crops of the Lowland Tropics*, ed. by C.L.A. Leakey and J.B. Wills, based on data from the Food and Nutrition Centre, Manila, Philippines.

So even in a diet which contains no animal protein at all, there are four main vegetable sources available:

a. legumes (peas and beans)
b. cereals
c. leafy green vegetables
d. some root crops (sweet potatoes, Irish potatoes, yams)

These four types of food, eaten together, can meet one's protein needs very satisfactorily, though each by itself is likely to be inadequate through lacking one or more of the wide variety of amino acids needed to build up the human body's protein requirement. Thus, in an area where grain crops are the staple food, beans and green vegetables from the garden should be eaten in the same meal as cereal foods to achieve a desirable protein balance. But where cassava, which contains very little protein, is the staple, it will be important to grow some cereal in the vegetable garden with a view to its use as a vegetable along with beans and green vegetables. The protein in yams and sweet potatoes will be more useful to the body if eaten with food from other protein sources such as greens or legumes.

An example of the benefits of combining different protein sources in this way is provided by some experiments, including feeding trials, which have been done in India. By adding a green leafy vegetable from the amaranth group to a diet consisting mainly of cereals and beans, the quality of the protein, and its utilization by the body, was improved so much that it was almost equivalent to the best animal protein, such as that contained in milk.

Cooking vegetables

It is important to note that new ideas about the cooking of vegetables may need to be introduced in parallel with new approaches to their cultivation, if the full nutritional benefit is to be achieved. Vitamins and minerals contained in crops can so easily be destroyed or lost through unsuitable cooking. Careful attention should be paid to traditional recipes and cooking methods.

In addition, it is important that food should be palatable, and indeed, enjoyable to eat, so it is right to grow some vegetables for the sake of flavouring and seasoning food – including onions, garlic, chillies, and peppers, according to local taste. These crops do also contribute a little in terms of minerals and vitamins.

Some general principles to observe in cooking are as follows:

a. Eat vegetables as soon after harvesting as possible – some of their vitamin content deteriorates in storage. Store leafy green vegetables in a dark place.
b. Do not add bicarbonate of soda, which reacts with and destroys some vitamins (vitamin C especially).
c. Use a little vegetable oil when cooking carrots or greens, and eat salads with oil – it helps the body to absorb the carotene (pro-vitamin A) from these plants.
d. Except when cooking cassava, or other vegetables containing inimicable substances, keep cooking times as brief as possible – many vegetables can be eaten raw, though they are better lightly cooked.
e. Use as little water as possible in cooking green vegetables, and always use the

water afterwards for drinking or in soups and stews – it contains some of the iron, vitamin C, and various salts from the vegetables.

f. When cooking dry peas and beans, blanch them at the outset by plunging them into boiling water and leaving them to boil for a minute or two – this softens the skin. Then leave them to soak in the usual way. Finally simmer them, but note that the initial blanching should allow the cooking time to be reduced. Do not add salt until the very end.

g. Where fuel for cooking is short, many vegetables are palatable after par-boiling, and chopped vegetables may be quickly fried in oil.

The eating of raw vegetables as a salad should not usually be encouraged, because of the difficulty of achieving adequate hygiene. Most regions which urgently need vegetable-growing programmes also lack modern water supplies. Salad vegetables are not likely to be adequately washed, therefore, and may have to be washed in water which itself is not safe to drink. Where sanitation is lacking, the vegetables may also carry hookworm from the soil. Thus the consumption of salads tends to multiply the health risks arising from bad water supply and sanitation. Fruit protected by a peel may safely be eaten raw with fairly minimal hygiene precautions, but otherwise, quick cooking methods (par-boiling and frying) have the advantage of economy in fuel and of conserving the vitamin content of the food.

5. Agronomy of Vegetables Suitable for Gardening Projects

(A brief review of some of the more important species; for more detail, see the books by G.A.C. Herklots and H.D. Tindall listed in the bibliography.)

Climatic Factors

Climate as well as the cultural and social factors stressed previously will always tend to favour local vegetables, which will be adapted to prevailing conditions of rainfall and temperature. However, for many of the more common vegetables, varieties suited to a very wide range of climatic conditions are available, and it is advisable to enquire about this when starting a garden. It is important to obtain good quality seed – many disappointments result from failure to do this – and it is also important to ensure that the seed obtained is for a variety suited to local conditions. Advice should be sought from government agricultural staff, agricultural institutes and research stations, and reputable seed merchants in the country concerned.

Vegetables transferred from temperate zones to the tropics are not only badly adapted to the heat of tropical climates, but may be adversely affected by the short day-lengths as compared with the long days of summer in northern latitudes.

Many common vegetables originated in Central and South America (potatoes, tomatoes, maize, chillies), but even so, they vary in their tolerance of heat. 'Irish' potatoes, for example, were originally grown in the cool parts of the high Andes, and are basically a highland crop when grown in the tropics today. Some vegetables do best in the cooler seasons of the year, and others, which may be adapted to high day temperatures (such as many local types of cabbage), prefer a climate where there is a marked difference in temperature between night and day. Such vegetables are most successful when grown at relatively high altitudes above sea level. These points are indicated in table 3, though again, it should be noted that many crop varieties have been produced for local conditions, and these cannot be shown separately here.

As an illustration of how altitude above sea level affects the choice of vegetables, it is worth comparing lists of crops recommended in one book (by Milsum and Grist – see bibliography) for different areas within the same region, about 5° north of the equator. In highland areas (higher than 1500 metres above sea level), temperate vegetables could be grown, and cabbage, Irish potatoes, and garden peas were recommended, but in lowland areas the same authors advised more strictly tropical vegetables, such as amaranths, kang-kong, Ceylon spinach, okra, brinjals, yams, and sweet potatoes. Vegetables recommended equally for both highland and lowland gardens included carrots and tomatoes (which would probably not do too well in the hotter lowland areas), and also French beans (whose many varieties include some which are suitable for tropical conditions).

Wet rain-forest areas, with rainfall above 1800 mm (i.e. 70 inches), tend to be difficult for vegetable growing, though some amaranths and many tree crops (pawpaws, avocado) do well. In dry areas (rainfall less than 1200 mm or 50 inches), the season for growing vegetables will be very short unless irrigation is possible.

For the type of garden project discussed in this manual, there are several temperate vegetables which are *not* recommended, either because they are difficult to grow successfully in the tropics, or because there are tropical vegetables, often popular locally, which have similar nutritional value, and suit local food cultures better. These non-recommended vegetables include garden peas, broad beans, and the scarlet-flowered runner beans. Cabbage and lettuce do grow well in the tropics, but are not so suitable for nutrition-oriented gardening as the spinaches; they are also readily saleable, and may tempt gardeners to sell them rather than to use them in the family diet.

Spinaches and Amaranths

The term 'spinach' refers to a range of widely different plants which are used as green vegetables, and have similar taste and nutritional value. By growing two or three different kinds, it is often possible to arrange for a garden to be producing spinach continuously throughout the year.

Spinach beet (with entirely green leaves) with which is included seakale beet or Swiss chard (white-ribbed leaves) is the easiest and most productive type in all but the hottest lowland areas. It will grow in a wet climate, provided that the soil does not become water-logged, and it can resist a dry spell to give a crop when the drought is over. The seeds are sown direct in the garden, and not in a special seed-bed for transplanting later. Rows should be about 60 cm apart, and the crop should be thinned to a spacing of just over 15 cm within the rows. (These distances can be measured out easily in the garden by using finger measurements – see figure 1.) When the leaves of the spinach beet are large enough to eat, they are simply pulled off, a few at a time, and the beet produces more, providing a continuous supply for several months.

(continued on page 35)

Tomato plants on a demonstration plot in an African nutrition programme.
(Photo: Oxfam)

A stall selling vegetables produced by local gardeners to women visiting a nutrition education centre and demonstration garden. Good quality seed is also available from this stall at the appropriate season for sowing. This exceptionally well-constructed stall is at the Valley Trust site in Natal (Project RSA 16. Photo: Oxfam).

Nutrition education: mothers attending a clinic for young children at a hospital in Ethiopia take part in a cooking demonstration. ***(Photo: Oxfam)***

Digging hoes being used in Guatemala to form terraces on a hillside, so that crops can be grown without causing soil erosion. (Project GUA 1. Photo: ***Oxfam****)*

As part of an applied nutrition programme in Orissa, India, boys were taught vegetable gardening at school. These two boys were then provided with seed by their school master, and were very successful in applying what they had learned in their garden at home. (FAO photo by I.A. Simpson)

An agricultural demonstrator talks with a woman gardener on the plot where her crops of beans and maize are growing. (Project RSA 16; Photo: Oxfam)

Amaranths are among the most important traditional vegetables in the tropics. They consist of a group of closely related species whose botanical names are given in table 3, some are commonly known as 'African spinach' and 'Chinese spinach'. They are good vegetables to grow wherever the climate is too hot and wet for other leafy vegetables, and they are widely eaten by poor people in Africa, India, and South-East Asia. In Senegal, for example, three species of amaranth are regularly eaten, some of them being collected as wild vegetables while others are grown in gardens.

Table 3. Summary of vegetables mentioned in the text

3(A) Green leafy vegetables, important mainly for protein, pro-vitamin A, vitamin C, iron, and calcium

Common Name (as used in text) & botanical names	*Other names*	*Growing conditions (altitude and season)*
Spinach beet & seakale beet *Beta vulgaris*	Swiss chard (for seakale beet); Leaf beet	best above 500 m, grow in dry season with irrigation.
Amaranths:		
Chinese spinach *Amaranthus gangeticus (A. oleraceus)*	Ranga sak (India) Lal sak (India)	the various species and varieties of amaranth are ill-defined; most grow best below 500 m, typically at the start of the rainy season
African spinach *Amaranthus hybridus*	Bush greens Spinach greens	
Spineless amaranth *Amaranthus gracilis*	Kulitis	
Ceylon spinach *Basella rubra*	Pasali-kirai (Tamil); Put, purai (Assam)	below 500 m; best in rainy season.
Water spinach *Ipomoea aquatica*	Water convolvulus; Kang-kong (Malaysia); Vallai-kirai (Tamil)	below 500 m; requires wet conditions; some varieties do well in rice fields.
*Cassava leaves *Manihot esculenta*	Manioc; Tapioca	anywhere with temperatures regularly above 60°F
*Sweet potato leaves *Ipomoea batatas*	–	moist soils below 1000 m.
*Okra leaves *Hibiscus esculentus*	Lady's fingers; Gombo; Bele	dry and wet seasons below 1000 m.
*Melon leaves *Cucumis melo*	also leaves of other curcubits: pumpkins, squashes, etc.	
*Papaya leaves *Carica papaya*		
*Cashew nut leaves *Anacardium occidentale*		

*denotes examples of plants grown for other purposes whose leaves are often eaten as green vegetables.

3(b) Legume vegetables, important mainly for protein and B-vitamins, and sometimes for iron and calcium

Common Name (as used in text) & botanical names	*Other names*	*Growing conditions (altitude and season)*
Cow-peas *Vigna unguiculata*	Paythenkai (Tamil); Long beans; other names for different varieties	below 1000 m best; dry season or late in rainy season
Green-gram *Vigna mungo*	Golden-gram; Mung bean; Tientsin green bean	best in the dry season below 500 m; tolerant of high temperatures but not of high humidity.
French beans *Phaseolus vulgaris*	Kidney beans; Haricot beans; String beans (names refer to different varieties)	very tolerant of all kinds of conditions, but best above 500 m; climbing varieties do best in high rainfall areas.
Lima beans *Phaseolus lunatus*	Seemai-motchai (Tamil); Butter beans	best at altitudes between 500 and 1500 m; do badly in very hot weather
Dhal *Cajanus cajan*	Dal; Pigeon pea; Red-gram	best below 2000 m, but tolerant of wide variety of conditions.
Groundnuts *Arachis hypogaea*	Peanuts; Monkey-nuts	varieties differ greatly in requirements. Frequent but light rainfall often crucial; sandy soils preferable
Soya beans *Glycine max*	Soy-beans	best below 1000 m; sensitive to rainfall and humidity in early stages.

3(c) Solanaceous crops and root crops; also maize

Common Name (as used in text) & botanical names	*Other names*	*Growing conditions (altitude and season)*
Solanaceous crops		
Tomatoes *Lycopersicon esculentum*	–	best above 200 m; dry season preferable, with irrigation; high humidity and temperature reduce yields.
Brinjals *Solanum melongena*	Aubergine, Egg-plant, Garden egg; Melongene; Kathiri-kai (Tamil); Baigan (Hindustani)	best below 1000 m; suited to dry or wet season, but avoid high soil temperatures and very wet soil.
Sweet pepper *Capsicum annuum* cv. *grossum*	Pimento; Guinea pepper	best below 1500 m; dry or wet season
Chilli pepper *Capsicum annuum* cv. *longum* (or cv. *acuminatum*)	Cayenne pepper	altitudes up to 1500 m; tolerant of high temperatures and a wide range of rainfall.
Irish potatoes *Solanum tuberosum*	Potatoes; Alu (Hindustani)	best above 1000 m; prone to disease in moist conditions so best in dry season.
Other roots and tubers		
Carrots *Daucus carota*	Gajar (Hindustani)	best above 500 m; typically do well in a sandy soil at the end of the wet season.
Sweet potatoes *Ipomoea batatas*	–	best below 1200 m; can be grown in wet or dry season provided soil is moist.
Cereals		
Maize *Zea mays*	Mealies (white grain varieties); Sweetcorn; corn (yellow grain)	best in the wet season at altitudes below 2000 m.

The amaranth known as Chinese spinach can grow into a very large, bushy plant, though it may be harvested before it reaches full size. It is easy to grow, and there are hardly any pests which attack it. The seeds are sown thinly in rows about 30 cm apart. The crop requires fertile soil to do well, and the first leaves may be ready for pulling in as little as 21 days. When the tops of leading shoots are removed at a first harvest, secondary branches develop, and a succession of harvests can be obtained. Leaves and flowers are cooked in soups and stews, or young leaves may be boiled and served like other green vegetables; they are rich in protein, calcium, iron, vitamin A and vitamin C.

Ceylon Spinach (not an amaranth) is another widely used vegetable in tropical Asia. It is a climbing plant which flourishes in the hottest and wettest climates. It is sown direct (i.e. not first in a seed bed) in rows about 50 cm apart, with a plant spacing of 20 cm, and stakes are provided for the plants to climb.

Water spinach, kang-kong, or water convolvulus is yet another plant whose leaves are eaten as a spinach, mainly in southern Asia. It is a semi-aquatic plant, grown near streams where there are high water tables, and sometimes grown in rice fields. Its leaves are rich in vitamins, minerals, and protein.

One problem with the spinaches and amaranths is that the leaves of many of them contain oxalates. These substances can be toxic if eaten in sufficient quantity, and can make some of the calcium and iron in these vegetables unavailable to the human body. Thus spinaches are in practice not as good a source of minerals as tables of nutritional values appear to suggest. The species which are worst in this respect are talinum (grown in the Caribbean area and West Africa), and one of the amaranths (African spinach). Thus, although the spinaches are of great value for their vitamin content, there are some species which are best not consumed in large quantities.

Solanaceous crops – Tomatoes, Brinjals and Peppers

Tomatoes, brinjals (egg-plant), and peppers, together with tobacco and potatoes, all belong to the same botanical family, the Solanaceae, and many of the same diseases attack all these crops. It is even said that a smoker tending tomato plants with unwashed hands can infect them with diseases carried by his pipe tobacco. The significance of this is that if the same plot or bed in a garden is used to grow tomatoes, brinjals, and then potatoes in succession, there is a danger that infections will build up in the soil. Thus the garden should be planned so that plants in this group are grown in a different bed each season, in a long rotation. Other precautions which can be taken to prevent the spread of disease are to burn all diseased plants, and to take great care in transplanting or pruning plants to limit broken stems or roots, because infections can enter such wounds. Some tomatoes produce better fruit if pruned, but if there is reason to fear disease in a particular case, it may be wiser to omit pruning altogether.

One other point that tomatoes, brinjals and peppers have in common is that they are all usually grown first in a seed-bed, or in boxes or pots. This is partly because the seedlings need to be protected from wind, heavy rain, and very hot sunlight. It is often possible to arrange a simple shelter over the seed-bed which will shield it from wind and rain, and which will allow sunlight to filter through the roof while giving some protection against the sun's full heat. A light roof of leaves or grass through which the light can be seen (not heavy thatch) can usually be improvised.

Tomatoes are grown in almost all tropical countries but do not thrive in the hot tropical lowlands, where they are specially prone to pests and diseases, and where the fruit may not set well. Growing them in these circumstances is a specialist job. In many highland areas, however, they are very suitable for household gardens.

Tomatoes are of great importance in improving the palatability of many dishes, but most of those actually eaten are of low nutritional value. There are many varieties of tomato, differing considerably in vitamin content, some, indeed, are rich in carotene (pro-vitamin A), but many contain relatively little. Plant breeders are able to produce improved varieties, increasing the content of carotene ten-fold, but this is at the expense of red pigment, and produces an orange-red fruit. Examples of varieties which *do* supply good quantities of carotene are Caro-Rich, Golden Jubilee, and other "tangerine" types, but the performance of these strains in the hottest tropical climates is not likely to be good.

Tomatoes are transplanted from pots or the seed-bed when the plants are about 10 cm high, and are spaced in rows about 40 cm apart. High soil temperatures do not favour the plant, and it is a good thing to provide a mulch to protect the ground surface from the heat of the sun; regular irrigation is desirable in dry weather, and the plants need the support of stakes as they grow.

Brinjals, also known as egg-plants, garden eggs, aubergines and melongenes belong to two main species, one of which is native to Africa, and the other to India. Brinjals are widely cultivated in India, and are said to be among the highest-yielding vegetable crops there. They will grow well at most altitudes except the highest, and in the wet or the dry season. The roots are sensitive to high soil temperatures, and therefore a mulch is needed; they also require well-drained soil, being adversely affected by water-logging.

Peppers are of two main kinds, sweet peppers and chillies, both being different varieties of the same species. The larger fruit of sweet peppers make a considerable nutritional contribution to many diets. The crop is tolerant of most tropical conditions, except excessively wet soil. The seedlings are transplanted when about 10 cm high. A spacing of about 40 cm between plants is desirable. The fruits are harvested while green, before they change to red.

Legume vegetables

Legumes or beans are of great importance in agriculture not only because of the protein they contain, but also because they are capable, in the right conditions, of improving soil fertility. Bacteria inside nodules on their roots take nitrogen from the air and 'fix' it in a form which the plant can use directly to build up its protein content – but much of this nitrogen may remain in the soil and can be used by other plants. The nutritional value of legumes is thus paralleled by their agricultural value, and an important technique in growing vegetables is to combine or rotate crops to take advantage of the fertility contributed to the soil by legume crops. But note that legumes do not always or automatically 'fix' nitrogen. If the bacteria are lacking or the soil temperature is too high, the legume crop may grow without any nitrogen being added to the soil by its root nodules. Sometimes, when the legume is new to an area, the seed may need to be innoculated with the relevant bacteria. Also, the crop should be mulched to keep the soil temperature down.

Almost all the individual gardening projects mentioned in sections 2 and 3 grow beans or other legumes regularly. One garden in India (KN 13) grows French beans; others in Africa and Bangladesh grow dhal (i.e. red gram or pigeon peas; ZAI 67, BD 70). Groundnuts are widely grown wherever there is concern about protein deficiencies in the diet (RSA 7). Lima beans and cow peas are widely recommended, and the four-angled or winged bean has been hailed as an under-exploited crop with great potential for the future.

Besides being a good source of protein – though not necessarily better in this respect than leafy green vegetables (see table 2) – most legumes also provide some of the calcium and iron needed, and also some B-vitamins. However, absorption of the iron in legumes by the body is often very inefficient. It should be noted that many legumes contain materials known as trypsin inhibitors and haemoglutins which prevent the human body making full use of the protein content of the crop; this problem may be overcome by eating beans green and unripe, or by lengthy cooking times.

Cow-peas or long beans, including the variety known as yard-long beans, are widely grown in the tropics. The pods may be picked and cooked while immature; or the seeds may be dried and used in stews; and sometimes the leaves are eaten as well. Cow-peas are tolerant of most tropical conditions, but are not recommended for altitudes above 1000 m. Dwarf varieties and climbing varieties exist; the dwarf cultivar may be most convenient for small gardens where it may be grown as an intercrop with maize or yams. Where no other crops are involved, plants may be spaced at 15 cm intervals along rows 50 cm apart. Climbing varieties need wider spacing and poles for support about 2 m tall.

Groundnuts are appropriate for home garden projects mainly when they are used directly by the householder in stews rather than being processed to extract oil or make flour. Eaten in stews, they provide protein, some calcium, iron and B-vitamins, and also much-needed fats. Groundnuts require a light sandy soil to do well, the many varieties available are suited to a wide range of climatic conditions in the tropics; many are drought resistant, but all do best with fairly frequent rain except when flowering and when fruit is forming. Typical spacing of plants in a bed would be about 30 cm, with rows 50 cm apart.

French beans, with their several varieties including kidney, haricot and string beans, are very versatile and may be grown in all but the most rainy areas in the tropics. For example, French beans are widely grown in Upper Volta, both for eating green and fresh and for harvest as mature dry beans. In the dry season, dwarf varieties are grown under irrigation; green pods are obtained 60-70 days after sowing with varieties known as Contender and Tenderlong. They are sown at 30 to 40 cm spacing, typically in rows spaced 30 cm apart. In the wet season, a climbing variety called Sossogbe is grown, which takes 80-90 days to maturity; it is sown at a more generous spacing to allow for staking later.

Soya beans must be mentioned here because, in the past, they have received so much emphasis in nutrition-oriented agriculture. They are a very rich source of protein and oils, and also a useful source of iron, which is absorbed by the body more easily than from most legumes. But despite its high nutritional value, soya is not really a suitable vegetable for small family gardens; it is fairly difficult to grow, and is often found to be unpalatable when used as a substitute for other beans. There is a good deal of confusion in some quarters about methods of cooking soya, so it may be useful to summarise the simplest methods:

a. when gathered green and unripe and cooked like green peas, soya beans are a very satisfactory vegetable.
b. soya beans can be used to produce bean sprouts.
c. mature, dry soya beans need to be soaked for a long time before cooking – up to 48 hours – and may then be cooked like other beans (e.g. haricot beans).
d. possibly the best way of using soya beans is to make soya milk following the traditional Asian method; it may be prepared at home in individual kitchens, but is probably best organised as a village industry – a "ten-women" business. Asian soya milk has a strong flavour which may not be acceptable everywhere, but a blander product may be produced by applying heat at the time when the beans are crushed. Soya milk is of great value as a sterile, nutritious drink in places where potable water is not available.

Other vegetables

Carrots are a temperate crop, unsuited to the high soil temperatures encountered in the tropics, but they are important as a rich source of carotene or pro-vitamin A, and they can be grown successfully in hot countries at altitudes above 500 m. Indeed, they are an important crop cultivated in various projects in India, Pakistan and Africa (KN 13, PK 42, RSA 16). They do best in light sandy soils with plenty of compost (but not animal manure). Seeds are sown thinly in rows 50 cm apart at the end of the wet season, and seedlings are thinned to a spacing of about 10-12 cm. The plants may be earthed up slightly as the roots begin to thicken to protect them from high soil temperatures. Regular irrigation is desirable.

Okra or Lady's fingers is an important crop in much of tropical Africa, India and the Caribbean. It is a tall, erect plant which produces pods containing up to 100 seeds in each. Young shoots and leaves are eaten; immature pods are cooked in soups and stews; leaves and pods are also dried for later use in stews. Okra is an important source of minerals and vitamins in some diets because of the quantity consumed – the leaves contain more vitamins than the pods. It grows in both the wet and dry seasons at altitudes below 1000 m; a suitable spacing is 40 cm between plants in rows 50 cm apart. Roselle is a related vegetable whose leaves are often eaten, and whose flowers are used in the preparation of beverages.

Bush Okra or Long-fruited Jute is an unrelated plant of superficially similar appearance whose leaves and shoots are eaten as a green vegetable; they are very rich in pro-vitamin A, calcium and iron. Bush Okra is easy to grow in the rainy tropics. Plants should be spaced about 20 cm apart with a similar space between rows.

Melons, Pumpkins and other curcubits are widely grown in Africa and will undoubtedly find a place in many gardens. The leaves are often eaten and are rich in vitamins. But the fruits, which in many places are harvested in large quantities and stored for eating over several months are of low nutritional value, apart from those types with orange flesh, which contain carotene (pro-vitamin A).

Onions, like the curcubits, are a widely grown and popular crop, but have relatively little nutritional value, though their usefulness in improving the palatability of other food should not be under-estimated. In some countries, notably Niger, onions are in fact the principal vegetable crop; they are usually grown in the cool, dry season of the year.

6. Problems and Methods in Starting New Gardens

Some common difficulties

Nutrition educationists who have tried to encourage vegetable growing frequently find it easy to interest people in the value of a vegetable garden, but find that this interest seldom gets translated into reality. Often there are practical obstacles, and as long as these remain unrecognised, nutrition and gardening talks will seem unrealistic to people listening.

Studies of several of the projects discussed here reveal that the obstacles most commonly found are problems with fencing and the availability of land. Then there are problems with obtaining seed of adequate quality, and with the work involved in running a garden, and finally, there is the experience of those who have already tried to start a garden and have lost all their vegetables because of a drought or through insect pests. It seems important that expatriates and other 'experts' who go around advising people to start gardens should first cultivate a garden of their own in the locality, so that they gain first-hand knowledge of the problems of drought, marauding animals, insect pests, and seed availability which are likely to be encountered. In the context of this booklet, it seems worth reviewing some of these particular problems:

a) *Fences* are usually essential, since vegetable-growing is impracticable without something to keep out cattle, pigs and other livestock; goats and monkeys are the most difficult animals to exclude. Where there is plenty of timber and sticks available, a pallisade can be constructed from these materials, but frequently, chicken wire and/or barbed wire secured on stout posts is necessary (depending on the local animals). This is expensive, and organizations promoting gardening should expect to help with costs, possibly through a revolving loan fund. If several gardens can be enclosed by one fence, this saves money.

b) *Availability of land* is often also an obstacle. Where there is no land at all which can be used, little can be done, but often the problem is that the land which might be used seems very poor, or is a long way from any source of water for watering vegetables. The solution may be for the project to start a communal garden or allotments, as discussed at length in section 3. Poor soils can sometimes be improved by means of compost or manure, and double digging will reduce compaction and improve drainage on neglected land.

c) *At first discouraging experiences* in growing vegetables may cause individuals to abandon a new garden at an early stage, and other people seeing this will then feel that starting a garden is not worth the effort. To guard against this, extension programmes should concentrate on vegetables that are easy to grow, such as a local spinach or amaranth, or beans of a variety known to grow well in the area; exotic vegetables should be avoided. The particular choice of vegetables should also depend on the time of year when the garden is being started, people have often had disappointments through planting what the nutritionist advised only to realise when the crop comes to nothing that it was the wrong season. Advice given should include seasonal variations with suggestions about the most suitable vegetables for the next planting period so that initial enthusiasm is not blunted. Where diseases and insect pests are the problem, it is best to consult local agricultural advisory services.

d) *Seeds* may be expensive, and the seeds sold in local stores may be of poor quality. Here again, those who promote the growing of vegetables have a responsibility to ensure that good seed is available at low cost, using bulk purchases from reputable seedmen where appropriate to get the best combination of reliable quality and low price.

A more general problem encountered in encouraging people to grow vegetables may be the impression that a great deal of manual work is needed. In fact, an established vegetable garden needs regular and frequent attention, but if it is well looked after, the work need be neither excessively time-consuming nor too laborious. Thus in communities where growing vegetables is seen as the women's role, it is not usually too difficult for the work to be fitted in with more strictly domestic tasks. However, starting a new garden may involve a considerable amount of heavy work and some expense. Fencing, clearing the ground, and digging it for the first time are all heavy jobs, which a woman with children may find difficult unless she has help from her husband. Where the project includes a component of community development or social education, so that it is seen as a matter for the whole community, not just for individual mothers, it may be easier to involve husbands or to get families to help one another (BRZ 165). In other instances, if the extensionists are men rather than women, they may be more effective in mobilising husbands to help (though this has to be weighed against the other advantages of having women extensionists when dealing mainly with women gardeners). In one or two areas where there are many mothers who are widowed or have husbands working away from home, the extension service has itself organized a labour force to set up individual gardens (RSA 16), with those helped paying for the labour (and any fencing or other materials) through a revolving loan scheme.

The latter approach may have more to commend it than might at first sight appear, because one of the best ways to learn a new technique is to work alongside somebody who is already skilled at it. Starting a new garden involves care in creating the right soil conditions, and the extension service should ideally be on hand with suggestions in any case. If advice is followed up by practical help in putting it into effect, with an experienced person working *with* the prospective gardener but not *for* her, then this may prove to be a very effective form of technical assistance 'at the grass roots'.

Steps in setting up a garden

The essential steps which individuals or groups need to take in setting up their garden can be summarised as (1) choosing the site, (2) fencing it, (3) acquiring gardening tools, (4) clearing the site, and (5) preparing the ground.

Many householders will not have a choice to make about the site for their gardens, but will have to use whatever land is attached to their houses. This may range from a tiny back-yard to a share in a large area of communally-owned land, where it may be necessary to negotiate permission to fence the garden plots. Where a community garden or allotments are planned, the project organizers should ensure that land is leased on a secure basis.

The size of the garden will depend on what land is available and how much time the family has to work it. In a moist climate (or in a dry climate with irrigation), 200 square metres (i.e. one fiftieth of a hectare) can provide all the vegetables needed by a small household throughout the year. Most gardens, however, may be smaller than this.

The ideal site is a level piece of ground with good soil near to a source of water (but not so near as to become water-logged in wet weather). If there is no water near the site, and it cannot be provided, the productivity of the garden will be greatly reduced, especially in dry areas, where cropping will only be possible during the rainy season. Gardens must therefore be planned in relation to the available water, and to minimise the amount of watering that has to be done. It is also important to use gardening techniques which conserve water and prevent its loss from the soil. These techniques include methods designed to maintain a high level of organic matter in the soil, especially through the use of compost, and mulching to keep soil temperatures down and prevent evaporation from the soil surface.

In one project (RSA 16), some small dams were built in conjunction with the larger gardens and provided a source of water. The opportunity was also taken to stock the dams with fish, so adding to supplies of protein, and linking animal production to gardening in an intelligent way.

In another gardening scheme where water was in very short supply (PK 42), small tanks were made, sunk into the ground, to collect water from the concrete slabs on which people did their laundry. Thus water carried to the village for washing was re-used on the gardens. More often, however, when there is no stream or other natural source of water, rainwater catchment tanks, or tanks collecting water from house roofs, may prove to be suitable water sources for gardens.

Gardening tools

In many parts of the world, some kind of digging hoe may be the principal or only gardening tool, used for preparing the ground, making seed-beds, weeding, and making ridges. There are many varieties with different shapes and weights of blade according to local custom, and local soil conditions. It is usually best to stick to the kind people are used to. The hoes illustrated in figure 1, for example, would not be much liked in some countries. Bigger hoes are used in Kenya, including a four-pronged fork hoe which is very useful for breaking and clearing the ground.

The Western-style spade and fork are not necessarily more efficient than the traditional hoe, and have the disadvantage that they can only be used in areas where people wear stout shoes or boots. However, some kinds of deep digging can be more readily carried out using a spade and fork than with a hoe. The Western-style hoes illustrated in figure 1 are adapted to a particular technique of weeding between growing plants while the weeds are very small. Traditional digging hoes are too heavy for this kind of weeding, and have their blades set at the wrong angle; if used for weeding they tend to dig too deeply into the soil and damage the crop.

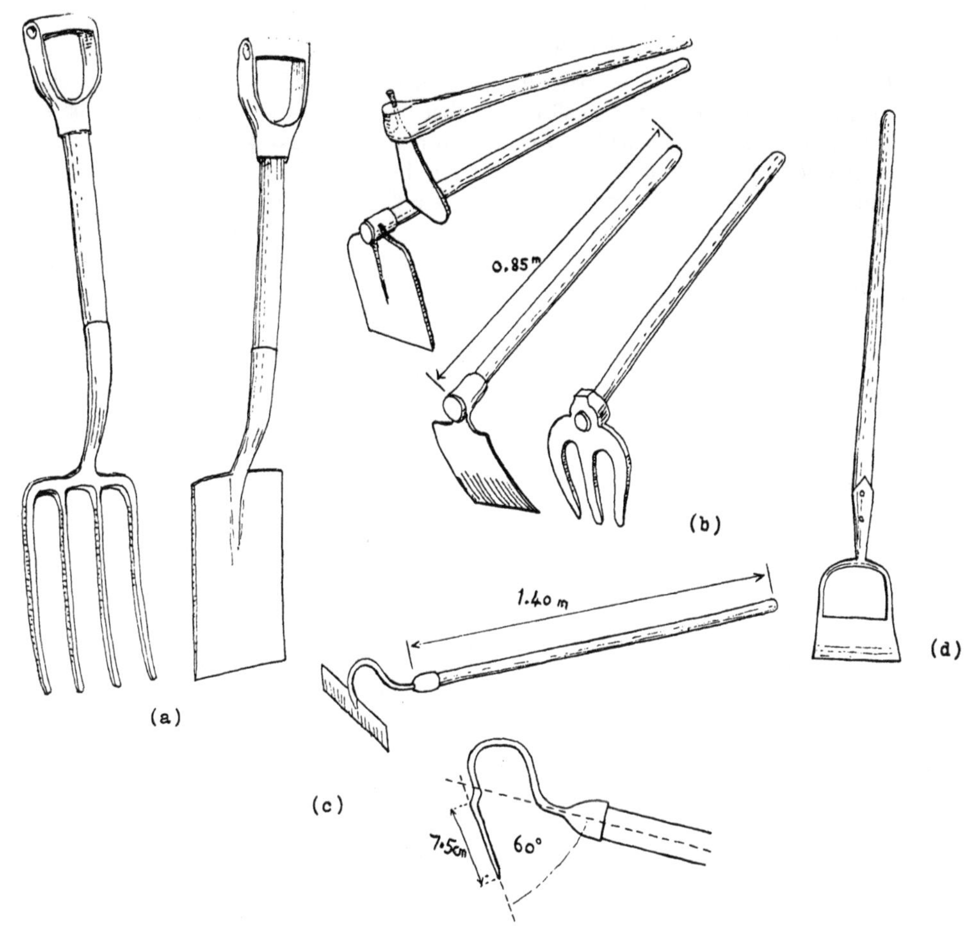

Figure 1. Digging tools (top) and weeding tools (bottom).
(a) Spade and fork for use by people who wear stout shoes or boots.
(b) Digging hoes and forked hoe as traditionally used in many countries; hoe blades vary greatly in shape from one region to another.
(c) Pulling hoe for weeding; note the light blade and long handle compared with the digging hoe; note also that commercially available types rarely have the blade at the optimum angle, and may need to be adjusted by careful bending.
(d) Dutch hoe, or pushing hoe, also used for weeding; this does the same job as the pulling hoe and some gardeners may find it easier to handle.

Spades are sold with either T-shaped handles, or handles like a D on its side (figure 1). The D-handle is more comfortable and efficient to use. It should be noted that imported tools are made in sizes to suit European men. Women may do better with shorter handles. Tools of local manufacture, made to suit local people's preferences, and local soil conditions, are greatly to be preferred (see photograph, page 32).

Clearing the plot and digging

Once a plot of ground has been fenced to make a garden, the first step is to cut down the grass, weeds and small bushes that may be growing on the site. Most of this material should be kept in a neat heap ready for making compost, but plants with woody stems should be separated from the rest, because they take a very long time to rot down into compost. At one time gardening books recommended that trees should be cleared from the site, because vegetables do not grow well in the shade, or where the roots are taking nutrients from the soil. In many areas, though, deforestation is doing so much damage that it may be best to keep as many trees as possible, planning the garden around them. Compost heaps and seed beds need some protection from too much sun, and it may be possible to reserve shady corners of the garden for these, though they should not be directly underneath trees. Where there are no trees on the garden site, land which slopes too steeply to cultivate and any awkward corners should be reserved for planting trees with edible leaves or fruit.

When the site has been cleared, a decision needs to be made about the layout of the garden. It is a good idea to have long, narrow beds for growing vegetables with paths in between, so that the gardener can reach the whole of the bed from the paths without treading on the cultivated soil. A common arrangement is to have beds about 120 cm wide (two short paces), with paths 50 cm wide in between. To prevent soil erosion, lay out the beds across the slope (i.e. along the contour), with the paths dug out a little to form gutters to trap any eroded soil. But to stop the paths from turning into water courses during heavy rain, a small bank should be constructed every three metres or so across their length.

Although this kind of layout may be regarded as widely applicable in the tropics, it should be modified in the light of traditional gardening methods in the area, the main proviso being that paths functioning as gutters, or furrows in larger plots, will always be needed to control erosion, unless the land is level, or made up in level terraces.

When starting a garden, especially on land which has been neglected, it is advisable to dig deeply so as to loosen the soil to a depth of at least 50 cm, and to dig in as much organic matter as possible, including compost or manure if this is available, or any vegetation such as crop wastes, straw, grass, or a specially grown 'green manure' crop (usually one of the legumes). If green vegetation is used, the fertility of the soil will take some time to develop, because the vegetation will first need to decompose into compost and become 'humus'.

When digging deeply, do *not* turn over the soil; the lower layers of tropical soils are often very infertile. The object of deep digging is to loosen the soil in the zone penetrated by the roots of crops and improve aeration and drainage. Once the garden is established, deep digging will not often need to be repeated, because normal gardening practice will keep the soil in good condition through regular additions of compost, and through the rotation of crops with different rooting depths.

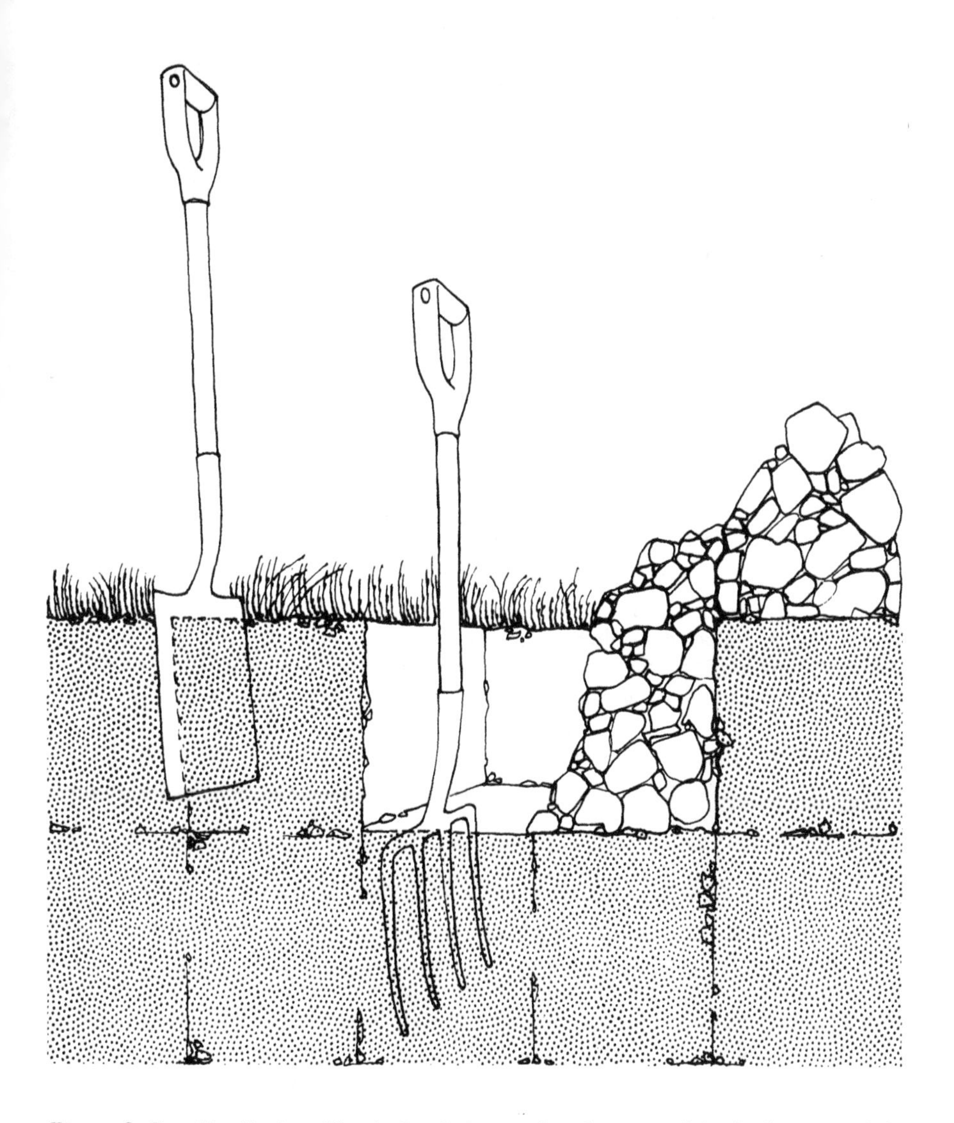

Figure 2. Double digging. The fork is being used to loosen soil in the bottom of the second trench or furrow, while the spade is placed ready to start a third trench.

One technique for loosening the soil to the required depth is *double digging,* which is most easily done with a Western-style spade and fork. The first step (figure 2) is to dig a trench at the end of the bed using a spade as deep as the length of its blade. A fork is then used to loosen the soil in the bottom of this trench to the depth of its prongs, but this bottom layer of soil is not lifted out or turned over. Then the spade is used to dig a second trench or furrow next to the first, filling in the original trench with the soil removed from the second one. The fork is again used to loosen the soil in the bottom of the trench; then a third trench is opened up and the process repeated. Where vegetation or manure is available to be dug in, this can be spread along the bottom of each trench before it is filled in with soil from the next one.

This technique is applicable only on the heavier types of soil; sandy soils will generally be sufficiently loose and well-drained to begin with, and do not need deep digging at all.

Digging should never be done when soils are very wet, as it can damage the soil structure; on the other hand, in the dry season, the ground is often too hard for digging to be practicable at all. The best time to be thinking of what crops to plant is at the start of the rainy season, and this is also the best time to dig. The soil should then be moist enough to be friable and easy to work, but not so wet as to create difficulties.

7. Gardening Techniques in Outline

The variability of geographical conditions

Having discussed the reasons for encouraging people to start vegetable gardens, and problems they may face in doing so, we must now consider the way in which the garden is managed – the routine of planting, weeding, mulching, tending, and finally harvesting crops, and the way in which one crop is planned to follow another in a carefully thought-out rotation.

Here it becomes even more difficult than in previous pages to provide simple guidelines which will be applicable in all tropical countries, because so much depends on climate, soils, the crops being grown, and local custom. All this booklet can do is to provide an introduction, and then to suggest that those who wish to promote vegetable-growing should seek as much information as they can locally, either from agricultural advisory services active in the region, or from gardening books written specifically for local conditions. The bibliography indicates some of the local gardening books which have been written for tropical areas in the past. Many of these are now out-of-date, and where they have not been replaced by more recent booklets or other materials, it may be worthwhile for an ambitious gardening extension programme to commission its own revision of earlier gardening booklets. In Zaire (ZAI 70), Oxfam has attempted to produce leaflets on gardening and agriculture written in very simple language so that local people with a basic knowledge of French can understand them; where local languages are used over wider areas, booklets should be produced in these as well. Similarly, the FAO "Better farming series" includes a booklet on market gardening in very basic

English, adapted from material originally produced in French for use in West Africa by the INADES organization.

Short booklets and leaflets of this kind are rarely sufficiently comprehensive to be used by gardeners as their sole guide, but they are useful in backing up advice given by extensionists, and in providing something for gardeners to refer to when the extensionist is not around.

All that is possible here is to outline the techniques which need to be covered in literature of this kind, and which need to be adapted to local conditions in every area where a gardening extension project is planned. These techniques can be listed as follows:

a. the uses of compost, manure and fertilizer.
b. making compost.
c. companion crops, mixed cropping and crop rotations; planning the sequence of crops to be grown.
d. tasks involved in growing a crop: sowing, transplanting, weeding, mulching, watering, staking, pruning.
e. pests and diseases.

The importance of compost

In approaching the subject of compost, manure and fertilizer, we turn from the nutrition of people to the nutrition of plants. There is, of course, a close relationship between the two, because the nutrients needed by a plant to sustain its life processes are also transformed by it into food which the human body can utilize; for example, the nitrate fertilizer applied to the soil by farmers is synthesised by the plant to create the proteins needed in human nutrition.

But in the same way as the theory of human nutrition has in the past been over-simplified (pages 22-23), so also there has been a comparable over-simplification of the theory of plant nutrition. This has led to reliance on chemical fertilizers, and a neglect of the importance of composts and manures. Chemical fertilizers supply many of the essential nutrients needed by growing plants, as do compost and manure, but these latter materials also provide organic matter, without which the conditions necessary for the plant to absorb nutrients efficiently may deteriorate.

It should always be emphasised that chemical fertilizers can never be a substitute for manure or compost because, although a fertilizer helps feed the plants, it does not do the other jobs done by organic matter in the soil, e.g. helping to keep nutrients in the soil for the plants to use and maintaining a good soil structure. When no manure or other organic matter is returned to the soil, and crops are grown with fertilizer as the only replenishment of soil nutrients, the soil is liable to deteriorate gradually until it becomes dust, very prone to erosion by wind or water, and a very poor medium in which to grow crops.

Fertilizer can be very beneficial when used in the right conditions, but the comments of Oxfam staff concerning two projects where chemical fertilizers are not used at all are worth noting. In Zaire, Oxfam agriculturists have felt that, away from the big plantations, "soil management is not at a level where the use of artificial fertilizer should be considered. For both crop and livestock production, the cost of imported machines, tools, and chemical products is prohibitive, and their availability is always uncertain. For the average Zairean farmer, his only reliable resources are his land, locally available seeds . . . and his own strength and intelligence. Making people use and safeguard their resources is the problem in agricultural development in the rural areas . . . " (ZAI 70).

Comment on another project which excluded the use of chemical fertilizers began by criticising this "organic only" approach, which often seems extreme and unbalanced. But the report went on to say that in the circumstances of Africa, "this philosophy for individual gardens and small cultivated arable areas is sound common sense. We have only to see the reduction in crops in Zambia and Tanzania to realise that a dependence on artificial fertilizers with their escalating price, moving them out of reach of the small man, is very dangerous." (RSA 16).

One reason why many farmers who can afford it opt for chemical fertilizer is that compost and manure are bulky materials and are difficult to handle, and compost in particular requires considerable care if it is to be satisfactorily prepared. However, these problems are much less for a gardener operating on a small scale, who should always have compost heaps in his garden, and should use manure whenever he can get it.

It is important to notice, though, that animal manure should not be applied to land with growing plants in the raw state, as it can 'scorch' the roots of plants. The best procedure is often to use the manure to make compost. Alternatively, it can be mixed with twice its volume of water, allowed to stand for a few days, and then be used to irrigate the garden. Poultry manure, however, may be dried, mixed with twice its volume of soil, then used as a top dressing.

How to make compost

Compost is likely to be the most important resource available to the woman gardener for maintaining the fertility of the soil, because she can prepare it herself using weeds and waste vegetation from her garden, grass cut from paths and unused corners, such animal manure as is available, and food wastes from the home. Vegetation and food wastes should be dried in the sun before being placed on the compost heap, and any woody vegetation should be separated from the rest. There are several ways of making compost; one which is quite widely used in the tropics can be broken down into the following steps:

a. Clear a level space sufficient for a heap about 1.7 metres square. Drive stakes in at the corners, and extra stakes along the sides to support the heap.
b. Lay down a layer of vegetable materials (grass, leaves, straw, preferably chopped into short lengths). This layer should be about 20 cm thick. On top of it spread a thinner layer (2.5 cm) of manure or animal excreta. Then sprinkle a little wood ash, and a little rich, fertile soil on top of the manure. Add a little ground limestone, too, if possible. Sprinkle a little water on each layer.
c. Add more vegetation, more manure, and more ashes in layers as in stage (b), and go on repeating the process until the heap is as high as it is wide (figure 3). Finish it off with a final layer of loose grass or straw.
d. Push sharpened stakes into the sides of the pile at various points, forcing them

Figure 3. A booklet on making compost written in Bangladesh includes a similar illustration of a garden with a compost heap being built up within a simple bamboo frame, which helps to maintain its shape. Material for a second batch of compost is being piled up in the frame alongside. (From IVS Package Program, Ambarkhana, Sylhet, Technical Bulletin No.8, How to make Fertilizer).

into the centre, and leave them in place. They are used for feeling the heat and wetness of the heap, and when they are withdrawn, they leave holes through which air can penetrate the heap.

e. The heap should be kept moist (like a damp towel) but not soggy. Cover it so that the sun does not dry it out too much, and also so that rain does not make it too wet.

f. After four or five days, remove one of the stakes from the centre of the pile. If the stake feels hot and slightly damp, the bacterial action which produces compost is working properly; if the heap is too wet, the stick will feel wet rather than damp. But if the heap seems too dry, sprinkle water on it.

g. After about 14 days, the compost heap must be turned. This means rebuilding it layer by layer to form a new heap next to the old one. The purpose of this is to mix the materials and allow the air to get at them.

h. If the pile continues to heat well, it can be turned a second time after another 10 days. Then after leaving the heap for another two months, making a total of about three months from the start, the compost should be ready for use – it should then have a sweet, earthy smell, and crumble easily between the fingers.

If the compost has not formed properly after three months, the heap should be turned again and left for a few more weeks. It should be understood that the process may not work if the compost heap is significantly smaller than the dimensions given, because the heap will not heat up sufficiently. If only a small quantity of material is available for making compost, the layers should be built up inside a wooden box, open at top and bottom, but with a light cover to keep out rain at the top. Another approach if only a small quantity of vegetation is available for compost making is to return the material to the soil as a mulch rather than make compost with it.

One way of using compost is to mix it with the top-soil at the time when the plot is being prepared for planting. Alternatively, it can be used as a top dressing around growing plants. It can also be used when transplanting seedlings; one or two handfuls of compost should be placed in the hole in which the seedling will be planted.

Planning the sequence of crops to be grown

In section 6 it was suggested that the garden should be laid out in narrow beds with paths in between. The arrangement of different vegetable crops in these beds will depend on which of several practices are followed. The aim should be to have several kinds of plants using the same soil, usually in a planned sequence, but sometimes by having two or more crops growing together. This makes it possible for the full potential of the soil to be exploited by crops which complement one another. For example, deep-rooting and shallow-rooting crops grown together will use the full depth of the soil; or legumes, whose net effect may be to add nitrogen to the soil, will usefully complement crops which need a lot of nitrogen.

The planned sequence of crops should also be designed to ensure that similar crops do not follow one another on the same ground in close succession – this reduces the danger of pests and diseases building up in the soil.

The conventional way of achieving this result is to use the principle of the *rotation of crops*. But in much traditional agriculture, many of the same needs are fulfilled by planting two or more crops together in the same bed – a system which is called 'mixed cropping', 'intercropping' or 'companion planting'. Since the companion crops may be planted at different times and harvested in different seasons, cropping sequences can be worked out which protect the soil from erosion by ensuring that it is always partially covered by a growing crop.

An optimum system is probably one which combines elements of both methods, including companion crops within a long rotation which ensures that no crop is grown on the same ground more often than once in two or three years. Whatever system is chosen, it will be wise first to consider traditional practice in the region concerned, and to enquire what companion crops or rotations are used. There will often be good reasons for them, and the practices should be used or adapted if at all practicable.

It is impossible to generalise about specific crop rotations because so much depends on local conditions. A typical plan may be based on a division of the garden into four, five or six separate plots, and the sequence of crops to be grown in one particular plot will then be set out as follows:

First year	January – April	tomatoes
	June – September	French beans
	September – January	carrots
Second year	January – April	lettuce
	June – September	onions
	September – January	maize

This is an actual crop rotation used in a part of the Caribbean where, with some irrigation, any time of year is suitable for most crops except late April and May. With a garden divided into six plots, each plot can follow the same sequence, but starting at a different point in the rotation. Thus, at any given time, all six crops will be growing. The situation becomes more complicated when particular crops have a strong seasonal preference, and can only fit into the rotation in, say, the January-April slot. It will also be obvious that not all these crops take the same length of time to grow to maturity. In instances where a crop is harvested some considerable time before the end of the allotted period, a quick-growing crop may be grown between the main crops in the rotation.

Rotations in other countries will vary considerably, according to what vegetables are being produced and local climatic conditions; it may also be desirable to include a fallow or rest period in the rotation when the ground is simply used to grow a cover crop or green manure which is later dug in. Some rotations grow beans more frequently than most other crops because of the demand for this type of crop and the benefit it may do the soil in the right conditions. But the general principles of all rotations are to separate solanaceous crops (tomatoes, brinjals, peppers, potatoes) because of their vulnerability to disease, and deep-rooting crops (e.g. maize and carrots) so that the deeper layers of soil are used at well spaced intervals.

In wet areas, or where there is water for irrigation, each bed should produce two or three crops per year if the rotation has been well planned. In a dry region, where only one crop per year is possible, planning rotation is considerably simpler although the garden will be much less productive. In dry areas, too, plants are more widely spaced, and it is sensible to grow crops such as radishes between the rows.

This use of quick crops is one kind of companion planting or mixed cropping, though the principle can be developed much further. The idea is to plant two or more crops together in alternate rows, or alternating within rows, choosing plants which will complement one another, e.g. in using the soil at different depths, or by arranging for a strong plant to shelter a weaker one, or by combinations which discourage pests. When a legume is one of the crops, its companion will benefit from the nitrogen which nodules on the legume roots put back into the soil. Experiments in India have shown that when beans and cereals are grown side by side in alternate rows, great economies are possible in the use of nitrogenous fertilizer without any reduction in cereal yields. Companion planting is traditionally practised in many African countries, and should not be discouraged by over-stressing the minor advantages of planting neat rows of one crop only. Indeed, one of the ways in which small-scale gardening can gain in productivity over larger scale operations is that companion planting is easily arranged, whereas in mechanised farming it is all but impossible. Examples of companion crops are:

Beans with maize,
Carrots with onions,
Maize with beans or pumpkins and melons,
Irish potatoes with beans or maize.

Observation of traditional practice in Nigeria shows that many mixtures of vegetables are grown, including okra with peppers, and also groundnuts with two other crops, okra and cow-peas. With the latter combination, the crops are sometimes grown in regular rows about 90 cm apart, and with plants placed at 25-30 cm intervals in each row. The sequence of plants in the rows is then, two groundnut plants, then one okra, then two more groundnut plants, then a cow-pea plant, then more groundnuts, and so on. In other villages where the same mixture of vegetables is grown, the groundnuts are first planted in rows 90 cm apart, then later, the okra and cow-peas are planted between the rows. (These plant spacings are not necessarily ideal; as often happens when plants are grown in rows, they tend to be too close to their neighbours within the row, while there is an unnecessary space between one row and the next. Planting companion crops between rows helps to overcome this disadvantage.)

It should be stressed that companion plants are not always sown or planted simultaneously. On Nigerian farms, in fields where millet and sorghum are grown, the system is first to plant an early-maturing bullrush millet as soon as the rains begin, but while they are still light and irregular. Two or three weeks later, when the rains are well established, sorghum is planted between the rows of millet. In another couple of months, the millet is harvested, the part-grown sorghum is weeded, and then cow-peas are planted between the sorghum plants.

In regions where this kind of practice is common and offers the advantage of using land more intensively, the rotation of crops in a vegetable garden will consist of an alternation of groups of companion plants rather than a long sequence of individual crops each grown by themselves.

Growing a crop: seed-bed preparation, sowing and transplanting

One of the differences between gardening in the tropics and in temperate zones is that when preparing the ground for sowing, the tropical gardener never breaks up all the lumps to reduce the soil to a uniform, fine tilth. Soil in this condition does not absorb rainwater as easily as a rougher surface, and is vulnerable to erosion.

The soil in a seed-bed should therefore be rough and coarse, to allow the rain to enter, except where seeds are sown. By rubbing one's foot along the line of the row, the soil can be made fine and firm for the seeds without breaking up lumps over the whole plot. Large seeds, like beans, are pressed down into the soil and then given a cover of 5 cm of loose soil. Smaller seeds are planted by scratching out a narrow, shallow depression along the length of the row (this scratch mark is called a 'drill'), the seeds are placed in the bottom of this and covered with 1-1½ cm of soil.

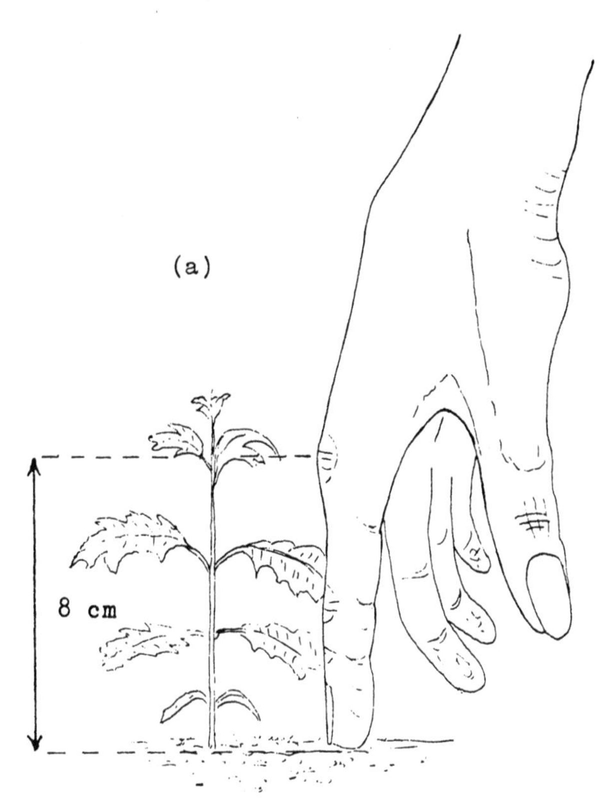

Figure 4. Finger measurements help to make sense of metric units and are convenient to use in the garden; the measurements shown are based on an 'average' woman's hand.
(a) Approximate size of tomato plants when ready for transplanting.

It is best to sow or plant in dry soil, and then water the seed-bed heavily after planting. Water seedlings every evening during dry weather. Within three or four days of the seedling showing through the soil, thin them out to a spacing of about 2 cm between each one, when they are a little larger, thin them out again to achieve the final spacing, which will depend on the type of vegetable concerned.

When seeds are first sown in boxes, or in a nursery bed, and the young plants are later transplanted into the vegetable plot, this should be done when they have five or six leaves, which will usually be about four weeks after sowing. Tomato plants are usually about 10 cm high when they reach this stage, i.e. they are a little taller than the length of one's index finger (figure 4). Transplanting should be preferably be done in the evening. The best way is first to saturate the seed-bed with water. This helps to ensure that the plants may be lifted with a good ball of earth around the roots. In the plot where the plants are to be grown, the soil is left in a

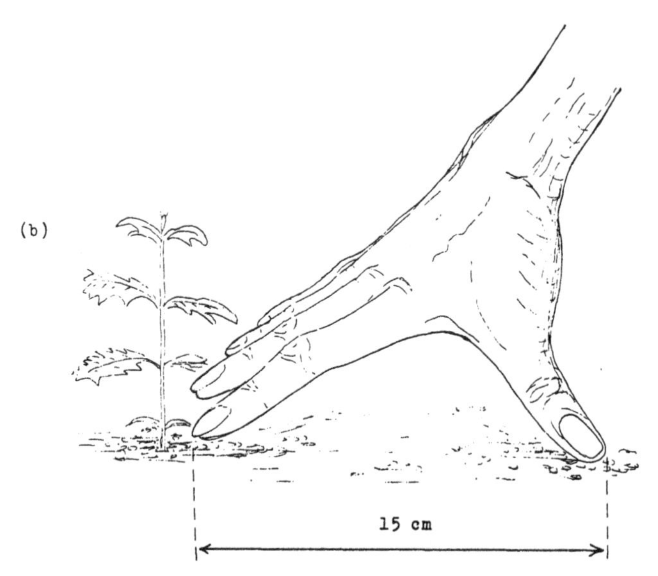

Figure 4b. Approximate spacing of spinach beet in the row should be slightly more than the distance shown; the spacing of tomato plants should be 2½ times this distance.

dry state at first, and holes are made to receive the plants with a pointed stick. A little compost is placed at the bottom of each hole, then the seedlings are placed in the holes, and the roots are buried up to the crown. Finally, the earth is pressed down firmly around the plants, and the bed is watered.

With traditional methods, gardeners use seeds saved from the previous year's harvest, or by leaving selected vegetables to go to seed. One aim of gardening schemes should be to make it possible for gardeners to buy seeds of improved varieties of vegetables. To ensure that quality is maintained, these should be bought regularly every year.

Among the vegetables most strongly recommended in this manual, the only ones which are normally grown in boxes or a nursery bed and then transplanted are the solanaceous crops (tomatoes, peppers). In places where the sun is very hot, or the rain is very heavy, the nursery bed should be protected by a low shelter with a light roof through which a little sunlight can filter (figure 5). Even with vegetables which are not usually transplanted, many will benefit if seedlings can be sheltered in some way from the full heat of the midday sun. This applies particularly to temperate vegetables such as spinach beet and Swiss chard.

Taking care of the vegetable crop: watering, weeding and mulching

It goes without saying that in dry spells, or during the dry season, a vegetable garden will need regular watering if it is to be productive. As a rough guide, a typical bed measuring 120 cm (i.e. 1.2 metres) across and 8 metres long will need about seven watering-cans full of water every day in the dry season (i.e. about 15-20 litres).

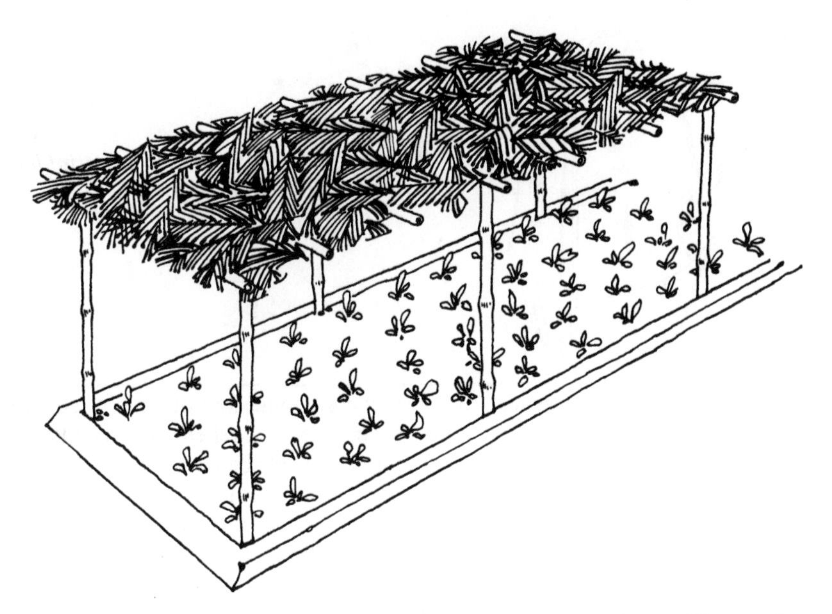

Figure 5. A simple shelter to protect seedlings against sun and rain. The roof may be about waist-height, and need not be thickly thatched – a little sunlight should be able to filter through it. (by permission of FAO).

The watering-can should have a rose with very fine holes (figure 6) so that the water will wet leaves and soil generally, and so that seedlings are not damaged by large drops. Watering a garden throughout a long dry season can mean a lot of work. Sometimes it may be possible to arrange a simple irrigation system, but if all the water has to be carried, it may be desirable to keep only a part of the garden in production during the dry season, so that all the beds are planted only when the rains come. Although a policy of this kind may be inescapable, it should be remembered that nutritional problems are often greater during the dry season, when local stocks of grain will be dwindling and food prices in the trading stores will tend to rise. Thus every effort should be made to make sure that water is available at vegetable plots, so that home-grown food can be produced at this nutritionally difficult time of year.

Watering a garden means that weeds will grow more rapidly as well as crops. Furthermore, heavy water drops can compact the soil, causing a hard surface layer to form. The weeding hoes shown in figure 1 are not only useful for removing the weeds while they are still small, but are well adapted for breaking up any surface layers that might have formed, so that soil can absorb water and be aerated more readily. These hoes are not intended to penetrate deeply into the soil, but are used to stir the surface soil and cut weed stems, being pushed (or pulled, depending on type) with light, short strokes.

In tropical gardens, many vegetables benefit from being earthed up as they grow, including tomatoes and beans as well as Irish potatoes. This can be done in stages

Figure 6. Watering can with its sprinkler rose. (by permission of FAO).

each time the garden is weeded. In areas where people do not have weeding hoes but only the heavier digging hoe (figure 1), earthing up is relied on to a considerable extent to bury weeds, and the weeding action just described is not used.

In gardens in the temperate zone, one problem is that the soil is often not warm enough, and seeds take a long time to germinate. In the tropics, however, it is desirable to protect the soil from the sun with a mulch to prevent it getting too warm. This is particularly important for legume crops, whose root nodules with their nitrogen-fixing bacteria do not function well in over-heated soil; many other crops benefit from mulching also, especially tomatoes and Irish potatoes.

Mulching means spreading material such as chopped straw, grass or leaves on top of the soil and around the plants. This not only keeps the soil cool, but conserves moisture, and discourages weeds by shutting out light from the soil. In the course of time, the mulch decomposes and benefits the soil in the same way as compost. Less hoeing is needed when mulching is practised because there are fewer weeds, and the dead vegetation on top of the soil gives some protection from heavy rain, reducing the risk of erosion, or of a hard surface layer forming. Where weeds are cut down with the hoe, they are left to die in the sun and they then form part of the mulch.

There are three times in the growing season when mulching may be appropriate: just after planting; prior to transplanting; and when the crop is well established. Irish potatoes are one crop which may be mulched after planting. If loose straw or grass is spread evenly over the ground after the seed potatoes are put in, they are protected from the heat of the sun and weeds are kept down, but the young potato plants have no difficulty in pushing their way through the mulch.

When tomatoes or peppers are transplanted, it is often considered to be good practice first to spread a mulch evenly over the ground in which they are to grow. Then holes are opened in the mulch and the seedlings are transplanted. However, the most common method, which may be used with all types of vegetables, is to spread the mulch on the ground between the plants after they have started to grow. This kind of mulch, when used with large vegetables, may be as much as 15 or 20 cm thick. After the crop has been harvested, the mulch should be worked into the soil, where it will provide nutrients for the next crop.

Other tasks which must be carried out while vegetable crops are growing include, for some crops, staking and pruning.

Figure 7. Pruning a tomato plant; the shoots blocked out in this diagram are the ones to be removed by pruning. (by permission of FAO).

Tomatoes and brinjals need pruning with particular care, left to themselves, they put out many branches, and then the fruits come late and are small. The purpose of pruning is to reduce the number of branches and to encourage large fruits to form more rapidly. Pruning tomatoes starts after the plant has two flowers and one leaf above the second flower. Then pruning is done by removing buds which show between the leaves and the stem (figure 7). The object is to keep only one or two main stems with their leaves and flowers. It is necessary to prune several times during the life of the plant, typically at intervals of about three weeks; procedures differ in detail for different varieties of tomato.

Tomatoes also need some support while they are growing; they need to be tied to a straight, rigid stake firmly embedded in the ground so they are not blown over by the wind. Climbing varieties of beans, often referred to as pole beans, need poles or sticks about 2 metres long – though dwarf beans can be grown without any staking. Where two rows of climbing beans are grown side by side, the sticks are embedded in the soil in each row, but then lean towards each other to form an inverted 'V', and are tied together at the top for support (figure 8).

Pests and Disease

There are many diseases, insect-pests and other forms of life which attack vegetable crops, and they are often difficult to deal with once they become established. If the use of chemical sprays or dusts to control them is contemplated, advice should be sought locally from the extension service about the identification of the pest concerned, and the correct chemical to use.

In many cases, people will not be able to afford chemical pesticides, will not have the skill to use them effectively or safely, and will find them difficult to

Figure 8. Dwarf and climbing varieties of French beans compared; the ridge of soil at the edge of the beds is intended to conserve water by preventing run-off when it rains. (by permission of FAO).

obtain. If chemicals are not used for any of these reasons, or because it is felt wiser not to on general environmental grounds, it must be accepted that some preventable losses due to pests and diseases will occur. However, there are still many precautions which can be taken to prevent damage to crops. The rotation of crops, as already mentioned, helps to prevent the build-up of disease-causing organisms in the soil. Another precaution is to burn all diseased or insect-infested plants rather than putting them on the compost heap – but remember that burning destroys material needed by the soil to keep it fertile, and never burn anything unnecessarily. A third precaution is to keep seeds where they cannot be infected, preferably in tightly closed tins or bottles, though if they have been saved from plants in the garden, they must be thoroughly dried first. A fourth precaution is to keep any place where produce from the garden is stored thoroughly clean. As shelves or storage cribs are emptied, they should be carefully swept out, and all leaves and other fragments from the crop should be burnt, because the eggs of insects are often lying dormant in such refuse.

Harvesting the vegetables

One of the advantages of growing vegetables for use by one's own family is that individual plants, pods or fruit can be harvested as they reach maturity, and one does not have to wait for a saleable quantity to be ready. With some crops, especially spinach beet, and many beans, it is possible to harvest a little every day, and spread the harvest over several weeks. With other crops, the harvest can be spread out over a longer period by sowing the crop in small batches with a week or two between each one. Even with these precautions, however, there will be periods when more of a particular vegetable is available than the family can eat. In such instances, it will be useful to preserve and store the produce. For example, in dry, sunny climates, it is comparatively easy to dry a wide range of vegetables in the sun, including green beans, carrots and tomatoes. Some beans are good to eat while they are green and immature, but if left a few weeks longer, can be harvested in the dry state and are then fairly easy to store.

There are a few simple and obvious precautions to observe in harvesting produce, like making sure that one's hands are clean, being careful not to damage fruits like brinjals or peppers or tear the leaves of spinach and other green vegetables. It is also best not to pick fruits when they are wet, otherwise they will quickly start to rot. Spoiled or damaged vegetables should be used first, with the damaged part being carefully cut out and discarded.

Conclusion

Comments such as these can only give the briefest of introductions to the problems and practice of gardening in the tropics. It is of the greatest importance to seek advice locally which will be directly relevant to local soils and climatic conditions, to the kinds of vegetables it is intended to grow, and to local techniques and traditions. Such advice will also be able to take into account the availability of seeds, fertilizers and pesticides in the area concerned. Where these materials are expensive, or a regular supply cannot be obtained, there will be local experience to call on concerning more immediately available resources.

What matters more, however, than the technical details of how to run a garden is the social and cultural factors which decide whether people will wish to initiate or improve vegetable gardens, and whether they can organise the necessary resources of labour and land. Once problems of this kind are overcome, and once the desire to grow vegetables and feed the family better exists, it is relatively easy to learn the techniques of doing so.

It is also important to understand the institutional arrangements which are needed if vegetable gardening is to be widely encouraged throughout a community. Whether the institution concerned is a hospital, a village committee, or a loosely organised group of extensionists, it must be able to approach and work with relevant members of the community (women in some places; whole families in others). It must also be able to provide practical help and technical assistance at certain critical points, such as in buying seed or erecting fencing.

It is both dangerous and difficult to lay down abstract principles about how to tackle these social, cultural and institutional issues. The approach adopted in this manual has therefore been simply to describe different ways in which people have tried to organise gardening and nutrition extension programmes. One hopes that these accounts of efforts made in Brazil and Bangladesh, Zaire and Zimbabwe, will stimulate the invention of locally appropriate strategies for attacking the problems of nutrition and food production elsewhere. None of the programmes described is an ideal one, fit to be copied in every detail, but all of them show something of what can be done to combat the malnutrition so widely found in the world today.

Bibliography

a) General books on tropical vegetables and gardening in the tropics

H.D. Tindall, *Commercial Vegetable Growing,* Oxford Tropical Handbooks, 1968.

G.A.C. Herklots, *Vegetables in South-East Asia,* Allen & Unwin, London, 1972, (useful as a general reference book on tropical vegetables, and not limited to S.E.Asia).

USDA, *Vegetable Gardening in the Tropics,* United States Department of Agriculture, Circular 32, Washington D.C.

C.L.A. Leakey and J.B. Wills, (eds.), *Food Crops of the Lowland Tropics,* Oxford University Press, 1977, (good chapters on vegetables and legumes, but more technical in style than other books listed here; contains much material relevant to West Africa).

A.V. and V.L. Gibberd, *A Gardening Notebook for the Tropics,* Longmans, London, 1953, (simple and clear, but heavily biased towards experience in Nigeria).

John Jeavons, *How to Grow More Vegetables,* Ecology Action of the Midpeninsula, Palo Alto, California, 1974, (stimulating, clear, and relevant to hot climates, though not to the moist tropics, but contains some odd ideas).

T.M. Greenshill, *Gardening in the Tropics,* Evans Bros., London, 1964.

Arthur Thomas, *Gardening in Hot Countries,* Faber, London, 1965, (mostly concerned with flowers and ornamental gardens, but has useful chapters on gardening techniques and vegetables; a book to read for pleasure rather than for practical advice).

World Neighbors, "Feeding the Soil", in *World Neighbors in Action,* volume 3, no.1E, Oklahoma City, c.1975, (a very useful how-to-do-it section in a newsletter packed with information on fertilizer, compost, and mulching; reprints available).

F.A.O., *The Sun-drying of Fruits and Vegetables,* c.1972, (short but practical booklet by Jackson and Mohammed, from Agricultural Services Division, F.A.O., 00100 Rome, Italy).

b) Regional gardening books and pamphlets

Caribbean region

USDA, *Vegetable Gardening in the Caribbean Area,* United States Department of Agriculture Handbook no.323, Washington DC, 1967.

R.C. Wood, *A Notebook of Tropical Agriculture,* College of Tropical Agriculture, Trinidad, 1957.

T.W.A. Carr, *Succession (Rotation) in Vegetable Growing,* Farmers Bulletin, Ministry of Agriculture, Lands and Fisheries, Trinidad and Tobago, 1968 (one of a series of short leaflets).

West Africa

F.A.O., *Market Gardening,* Better Farming Series No.17, Rome, 1970, (in basic English, summarising techniques described by extensionists or in radio broadcasts; originally published in French as *La Culture Maraîchère,* Institut africain pour le developpement economique et social – INADES – 1967).

H.D. Tindall, *Fruits and Vegetables in West Africa,* F.A.O., Rome, 1965.

G. Grubben, *L'horticulture en Côte d'Ivoire,* Royal Tropical Institute, Amsterdam.

A.V. and V.L. Gibberd, *A Gardening Notebook for the Tropics,* Longmans, London, 1953, (Nigerian bias).
T.A. Phillips, *An Agricultural Notebook for the Tropics,* Longmans, London, 1956, (Nigerian emphasis).

Zaire

Mike Brownbridge, *Oxfam Agricultural Leaflets,* Oxfam, Oxford, 1977, (in simple French, these leaflets are very brief; one covers vegetable gardens and another compost making).

East Africa

Horticultural Handbook, volume 1, 1966, Department of Agriculture, Nairobi.
C.D. Hemy, *Vegetable Notes,* Ministry of Agriculture Bulletin No.10, (1961), Dar-es-Salaam.
A.G.K. Will, "Notes on a small vegetable market garden trial in Uganda", *First East African Horticultural Symposium,* Technical Communication No.21, 1971, International Society for Horticultural Science, Boschstraat 4, The Hague, Netherlands.

South-East Asia

G.A.C. Herklots, *Vegetables in South-East Asia,* Allen & Unwin, London, 1972, (not a gardening manual, but an extensive catalogue of vegetables with many practical tips on growing them).
S. Satiadiredja, *Indonesische groenten,* Wolters, Jakarta, 1950, (Indonesian vegetables, in Dutch).
J.J. Milsum and D.H. Grist, *Vegetable Gardening in Malaya,* Department of Agriculture, Kuala Lumpur, 1941, (useful, detailed and thorough, though now very old and difficult to obtain).

Indian Sub-Continent

H.F. Macmillan, *Tropical Planting and Gardening,* Macmillan, London, 1954, (specially relevant for Sri Lanka).
K.S. Gopalswamiiengar, *Gardening in India,* Bangalore (no date), (by an experienced gardener, and possibly one of the best books).
Publications of the Indian Council of Agricultural Research, (ICAR), Krishi Bhavan, New Delhi, 110 001:
Introduction to Gardening, (current price, Rs.7/10);
Vegetables for Tropical Regions, (Rs. 3/60);
Manurial Requirements of Vegetables, (more technical than the others, Rs.5/80).
Sutton's Garden Guide, Sowing Tables, (a seed merchant's booklet giving very useful advice as to the best planting dates for vegetables in the different regions of India: Rs.2/75 from Sutton & Sons (India) Pvt.Ltd., 13D Russell Street, Calcutta, 700 071).
Agnes W. Harles, *Gardens in the Plains,* Oxford University Press, Bombay, Rs.30/0.
M.S. Randhawa, *Developing Village India,* Orient Longmans, Calcutta, 1951, (useful chapter on horticulture; stresses fruit trees).
Harlan H.D. Attfield, *How to make fertilizer,* 1977, IVS Package Program Bulletin No.8, Ambarkhana, Sylhet, Bangladesh; also available from Volunteers in Technical Assistance in the U.S.A., (well-illustrated leaflet on compost making).

c) Background information on nutrition and health aspects

Data on the nutritional values of specific vegetables are given in several of the books quoted above, and in most detail by H.D. Tindall, *Commercial Vegetable Growing,* Tropical Handbooks, Oxford, 1968.

H.A.P.C. Oomen and G.J.H. Grubben, *Tropical Leaf Vegetables in Human Nutrition,* Communication 69, Department of Agricultural Research, Royal Tropical Institute, (Koninklijk Instituut voor de Tropen), Amsterdam, Netherlands, 1977; (an *exceptionally useful book* on the traditional vegetables of the tropics and their nutritional value).

Michael Latham, *Human Nutrition in Tropical Africa,* F.A.O., Rome, 1965.

Elizabeth Stamp, (ed.), *Growing Out Of Poverty,* Oxford University Press, Oxford, 1977, (documents some instances where agricultural development has sprung from medical concern about poor nutrition).

Katherine Elliott, *Training of Auxiliaries in Health Care,* Intermediate Technology Publications, 1974.

www.ingramcontent.com/pod-product-compliance
Lightning Source LLC
Jackson TN
JSHW011942131224
75386JS00041B/1528

* 9 7 8 0 9 0 3 0 3 1 5 0 9 *